SCIENCE AND RELIGION IN QUEST OF TRUTH

Science and Religion in Quest of Truth

JOHN POLKINGHORNE

Yale UNIVERSITY PRESS

New Haven and London

Copyright © 2011 by Yale University.

Yale University Press books may be purchased in quantity for educational, business, or promotional use. For information, please e-mail sales.press@yale .edu (U.S. office).

Set in Janson type by Tseng Information Systems, Inc.
Printed in the United States of America.

A catalogue record of this book is available from the Library of Congress.

ISBN: 978-0-300-17478-6 (cloth)

This paper meets the requirements of ANSI/NISO z39.48-1992 (Permanence of Paper).

10 9 8 7 6 5 4 3 2 1

The first serious book I ever read about science and religion was Ian Barbour's Issues in Science and Religion. *He has been a leading pioneer in the modern discussion of these questions and it has been a privilege to read his many writings and to know him as a friend and colleague in the quest for truth. I dedicate this book to Ian, with my thanks and admiration.*

Contents

Abbreviations for Selected Writings
of John Polkinghorne   ix

Introduction   xi

ONE. Truth and Understanding   1
TWO. Some Lessons from History   26
THREE. Insights from Science   33
FOUR. Theology and Science in Interactive Context   69
FIVE. Motivated Christian Belief   116

Postscript: Other Faiths   131

Index   135

Abbreviations for Selected Writings of John Polkinghorne

BG   *Belief in God in an Age of Science*, Yale University Press, 1998.

BS   *Beyond Science*, Cambridge University Press, 1996.

ER   *Exploring Reality*, SPCK/Yale University Press, 2005.

ES   *Encountering Scripture*, SPCK, 2010.

FSU   *Faith, Science and Understanding*, SPCK/Yale University Press, 2000.

GH   *The God of Hope and the End of the World*, SPCK/Yale University Press, 2002.

OW   *One World*, 2nd edn, Templeton Foundation Press, 2007.

QPT   *Quantum Physics and Theology*, SPCK/Yale University Press, 2007.

RR   *Reason and Reality*, SPCK/Trinity Press International, 1991.

S as T   *Scientists as Theologians*, SPCK, 1996.

SC   *Science and Creation*, 2nd edn, Templeton Foundation Press, 2006.

SCB   *Science and Christian Belief*, SPCK, 1994; in North America, *The Faith of a Physicist*, Fortress Press, 1996.

SP   *Science and Providence*, 2nd edn, Templeton
        Foundation Press, 2005.
ST   *Science and the Trinity*, SPCK/Yale University Press,
        2004.
TCS   *Theology in the Context of Science*, SPCK/Yale
        University Press, 2008.

In the citations, numbers signify the relevant chapters.

Introduction

Much of my writing has been in the form of comparatively short books focused on a particular topic—creation, providence, eschatology and so on. This fitted well with my formation as a scientist, for whom it has been natural and congenial to address particular issues concisely and one by one. The strategy of writing short books has also enabled me from time to time to return to certain key issues as my thinking in these matters has developed. My Gifford Lectures[1] had a wider scope, but one that was defined by the clauses of the Nicene Creed, approached in the style of 'bottom-up thinking', so natural to the scientist, of motivating belief through consideration of the evidence offered by experience.

Although taken together these books cover a large range of topics in science and religion, none of them individually corresponds to the substantial volumes surveying within a single cover the wide scope of interaction between these two quests for truth, such as those that have been written by my

1. SCB.

colleagues Ian Barbour[2] and Arthur Peacocke.[3] The nearest I have got to giving a synoptic account has been the writing of two slim volumes, one addressed to the general reader in fairly informal terms[4] and the other more academic in character, but intended to serve simply as an introductory text-book summarising the thoughts of others active in the field as well as those of my own.[5] In addition, I have written a short comparative survey of how Barbour, Peacocke and I had approached certain key issues.[6]

The arrival of my eightieth birthday has prompted me to consider whether it might not be time to attempt a fuller survey of the issues of science and religion as I see them, set out within the covers of a single book. To do so in the detail comparable to that of the 350–450 page volumes of my colleagues would have involved a good deal of recycling of what I had already written in my more narrowly focused books. Moreover, some of that material is already conveniently available, collected together in a well-constructed Reader that Thomas Jay Oord has edited.[7] Accordingly, I have adopted an intermediate strategy in the present volume. It covers what I believe to be the critical issues in the contemporary vigorous dialogue between science and theology, concentrating on what I judge to be the essential insights that are relevant to the issues raised. References are given to my earlier writings where chapters can be found that afford a more detailed en-

2. I. G. Barbour, *Religion and Science*, SCM Press, 1998.

3. A. R. Peacocke, *Theology for a Scientific Age*, SCM Press, 1993.

4. J. C. Polkinghorne, *Quarks, Chaos and Christianity*, SPCK/Crossroad, 1994.

5. J. C. Polkinghorne, *Science and Theology*, SPCK/Fortress, 1998.

6. J. C. Polkinghorne, *Scientists as Theologians*, SPCK, 1996.

7. T. J. Oord (ed.), *The Polkinghorne Reader*, SPCK/Templeton Press, 2010.

gagement with particular issues, but I believe that on its own the present book lays out the core concepts that are fundamental in what is surely one of the most significant interdisciplinary interactions of our time, pursued in a wide quest for truthful understanding. The book concentrates on my own work, not at all because I think that I have said all that need be said, but simply because, after more than thirty years of reflection, this is what I have to offer.

John Polkinghorne
Queens' College
Cambridge

SCIENCE AND RELIGION IN QUEST OF TRUTH

CHAPTER ONE

Truth and Understanding

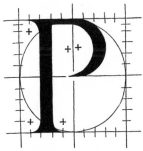EOPLE sometimes say that science deals with facts but that religion simply trades in opinions. In other words, science's concern is with truth, understood as correspondence with reality, but the best that can be said of religion is that it might be 'true' for an individual, but only in the weak sense that it was helpful for that particular person to look at life in that particular way, without necessarily implying anything about the way reality actually is. Two bad mistakes lie behind this claim.

The first is a mistake about science. There are no scientifically interesting facts that are not already interpreted facts. No doubt all could agree what the reading was on the dial of some piece of measuring apparatus, but for that reading to have meaning one would need to know what the instrument is actually capable of measuring. For that one needs a theoretical understanding of the nature and operation of the apparatus. In science, experimental 'fact' and theoretical 'opinion' intertwine in a subtle circularity, as experiment seeks to confirm or disconfirm theory and theory seeks to interpret experiment.

The second mistake is about religion. The question of truth is as central to its concern as it is in science. Religious belief can guide one in life or strengthen one at the approach of death, but unless it is actually true it can do neither of these things and so would amount to no more than an illusionary exercise in comforting fantasy.

Both science and religion are part of the great human quest for truthful understanding. Before we explore what this might imply for their mutual relationship, we must pay further attention to the individual characters of these two truth-seeking endeavours. The claim will be that both are seeking truth through the attainment of well-motivated beliefs.

THE NATURE OF SCIENCE[1]

Perhaps the first thing to say about science is that it has been wonderfully successful in its quest for understanding. Time and again it has been able to present results of the greatest interest which have been universally agreed by the whole scientific community. Repeatedly in science, questions actually get settled. At the beginning of the twentieth century, there were still some physicists who thought that the notion of atoms might be no more than a manner of speaking, useful for some purposes but not needing to be taken seriously as indicating the actual existence of a particle structure in matter. Today, it is universally acknowledged that matter has a granular nature, even if the current elementary constituents are quarks and gluons and electrons, rather than the much larger atoms. To

1. OW 2; BS 1–2.

take another example, when expounding his theory of evolution in 1859, Darwin had to appeal to the existence of unexplained small variations between the characteristics of successive generations of living beings, which produced results that could then be sifted and preserved by natural selection. We now know that these variations arise from mutations in the genetic material DNA. No other realm of human enquiry can present such a successively enlarging catalogue of successful agreed conclusions as that which science is able to claim.

Science has purchased this great success by the modesty of its ambition. It sets out to ask only the question of what are the processes by which things happen, bracketing out of its consideration other questions, such as whether there is meaning, value or purpose present in what is happening. Science is principally concerned to explore only one dimension of the human encounter with reality, essentially that which can be called impersonal, open to the unproblematic repetition of the same phenomena, irrespective of the place of investigation or the character of the investigator. Even in historical sciences such as cosmology or evolutionary biology, concerned with understanding unique sequences of events, much scientific explanatory power depends upon the insights of directly experimental sciences, such as physics and genetics. It is this self-defining limitation to impersonal experience that has given science the great secret weapon of experiment as its unique means for attaining intersubjective agreement. Repeatability of this kind is unattainable in any realm of personal encounter with reality, where each event possesses its own unique character, and the resulting diversity of experience makes complete agreement much more difficult to achieve. Science's declining to engage with the personal dimension of experience implies the limited character of the account that it can give of reality. A scientist, speaking as a scientist, can say no more about music than that it is vibrations in the air, but speaking as a person there would surely be much more to say about the mysterious way in which a temporal succession of sounds can give us access to a timeless realm of beauty.

Despite agreement being readily attained in science about immediately perceptible phenomena (all those watching see the pointer move to the same point on the scale), the question of the significance of the phenomena observed is made complex by there being the circular interaction, already noted, between experiment and theory in interpreting the meaning of the results. The frequent attainment of universal agreement in the scientific community arises from the conviction among scientists that this circularity is usually benign rather than vicious. A number of factors produce this belief.

A really successful scientific theory attains a persuasive naturalness of explanation from the fact that an economically formulated hypothesis is seen to lead, without forcing or manifest contrivance, to the understanding of a wide range of diverse phenomena. Darwin's theory of evolution not only made sense of the fossil record but also explained the existence of vestigial organs, such as the human appendix, and it made intelligible the local variations in species observed in groups of nearby islands, such as finches in the Galapagos Islands. Much later in the development of evolutionary biology, it became apparent that the order in which species had emerged could also be inferred from study of the differences between their genomes and the results of this independent assessment were found to be in good accord with the ordering derived from the fossil record, a compatibility that afforded an impressive confirmation of the basic concept of descent with modification.

A successful scientific idea frequently manifests long-term fruitfulness by showing a capacity to explain not only the phenomena that originally led to its discovery but also other phenomena, known at the time but not understood or taken into account in framing the theory. Even more impressively, the theory can also lead to the prediction of unanticipated phenomena which are subsequently found experimentally to occur. Paul Dirac discovered a celebrated equation that describes the electron. He hit upon it by finding an elegant way in which to combine quantum theory with special relativity. The equation immediately provided an unexpected bonus by turning out to explain a known, but till then not understood, aspect of the electron's magnetic properties. A little later the equation was shown by Dirac also to imply the existence of antimatter, a wholly unexpected consequence that was quickly confirmed experimentally. This experience of the long-term fertility of an insight strongly encourages scientists to take their discoveries with ontological seriousness. Unless there was a correspondence between ideas and reality, these successes would seem unintelligibly gratuitous. Instinctively scientists are philosophical realists, believing that what we come to know about the physical world is indeed telling us what that world is actually like.

Such a realist belief receives further support from the way in which the physical world is often found to resist our prior expectations and prove stranger in its character than we had thought, or perhaps even could have thought without being prompted by the stubborn nudge of nature. A striking example of this experience is provided by the discovery of quantum physics.[2] No one in 1899 could have supposed that light could manifest the oxymoronic property of sometimes behaving like a wave and sometimes like a particle. After all, waves are spread out and flappy, while particles are small and bullet-like. Nevertheless, as we all know, this is how light has actually been found to behave. This led eventually to the discovery of quantum theory, in which states can exist that are mixtures of possibilities that classical physics and commonsense would say could never be combined together (technically this is called the superposition principle). For example, in the clear Newtonian world there could only be states with a specific number of particles present in them (just look and count how many). However, in the cloudy quantum world there can be states that correspond to an *indefinite* number of particles (formed out of superpositions of states with different particle numbers). These are the states that turn out to display wave-like properties. The recalcitrant way in which nature can resist our prior expectation is a powerful incentive to believing that in science we are actually exploring a world that stands over against us in its independent character.

Nevertheless, the strangely elusive and counterintuitive character of the quantum world has encouraged some to suggest that the idea of entities like electrons which can be in unpicturable states such as superpositions of being 'here' and being 'there' is no more than a convenient manner of speaking which facilitates calculations, and that electrons themselves are not to be taken with ontological seriousness. The counterattack of the scientific realist appeals to intelligibility as the

2. See, for example, J. C. Polkinghorne, *Quantum Theory: A Very Short Introduction*, Oxford University Press, 2002.

key to reality. It is precisely because the assumption of the existence of electrons allows us to understand a vast range of directly accessible phenomena—such as the periodic table in chemistry, the phenomenon of superconductivity at low temperatures and the behaviour of devices such as the laser—that we take their existence seriously.

Belief in scientific realism is well-motivated, but one cannot claim that it is logically proved to be true beyond any possibility of question, as if it would be wilfully stupid for anyone to deny it. This relatively modest assertion of status of the belief recognises that there are some possible difficulties opposing a realist point of view, which now need to be considered and evaluated. The progress of science, with the changes of understanding that can result from this, make it clear that scientific achievement cannot be claimed to constitute the attainment of complete and absolute truth. Instead, science's exploration of reality must be seen as resulting in the creation of 'maps' of the physical world which are indeed reliable, but only on a particular scale. No map can reproduce all the detail of the terrain and changing the scale can lead to the exhibition of new features not previously recorded. The immense success of Newtonian physics had eventually to be qualified by the recognition that understanding phenomena on the subatomic scale required the quite different insights of quantum theory, together with the recognition that phenomena involving particles moving with velocities close to the speed of light required the insights of relativity. The Newtonian map was not torn up, but its limitations had been identified. Some philosophers of science, such as Thomas Kuhn, saw these changes as revolutions that subverted the realist claims of science. However, the issue is more subtle than that and Kuhn's conclusion does not follow. A really successful scientific theory, such as Newtonian mechanics, never totally disappears. In fact, it is still good enough to send a space probe to Mars. What happened was that the domain of Newtonian applicability had been circumscribed and the appropriate scale of its map had been determined. When a new theory, such as quantum theory or relativity, has been discovered, one of the vital tasks is to establish what are called 'correspondence principles', explaining how the new theory can attach to itself the undoubted successes of the old one in appropriate circumstances. The different maps of the reality offered by physics are not identical, but they can be shown to be mutually compatible where there is an overlap between them. While the achievement of science does not amount to absolute and exhaustive truth, it can be asserted to be what one may call 'verisimilitude', an ever tightening, but never total, grasp of physical reality. Science can claim to attain the discovery of yet more satisfactory levels of understanding, adequate to what is currently known, without pretending to rule out the possibility of future discoveries revealing an even deeper and more complex order present in the physical world. Its achievement can be characterised as a kind of convergent realism.

Michael Polanyi was a philosopher of science who brought to his task the prior experience of a long and distinguished career as a physical chemist. In *Personal Knowledge*[3] he recognises that there is no coercively logical certainty in science, yet he also maintains that it affords an understanding that should rightly command our intellectual assent and com-

3. M. Polanyi, *Personal Knowledge*, Routledge and Kegan Paul, 1958.

mitment. Not only does Polanyi duly recognise the delicately circular nature of the interaction of theory and experiment but also he identifies the need for acts of personal judgement in the practice of science, involving decisions taken with universal intent, that are open to assessment within the truth-seeking community of science but which are not simply the result of following an explicitly prescribed protocol. For example, all experimental analysis has to deal with the problem of 'background', that is, the possible presence of spurious effects arising from uncontrolled environmental influences, such as the collisions of stray cosmic rays accidently traversing a bubble chamber. These have to be eliminated or allowed for in some way. There is no little black book or computer program guaranteed to tell the experimenter exactly how to do this. Solving the background problem requires individual acts of personal judgement. These require the exercise of tacit skills—'we know more than we can tell' is a favourite Polanyi remark—that have to be acquired through apprenticeship within the practice of the truth-seeking scientific community. Polanyi tells us that he wrote *Personal Knowledge* to show how he could rationally commit himself to what he believed scientifically to be true, although he knew that it might be false.

The concept of commitment is very important in Polanyi's thinking about the nature of science. People sometimes say that scientists doubt everything, but in fact that would be a stultifying policy to pursue, leading to a paralysing enslavement to uncertainty. Instead, scientific discovery requires the boldness of provisional commitment to a point of view, while remaining aware that this may require subsequent modification in the light of further experience. Above all, science requires commitment to the basic act of faith that there is a deep rational order in the world awaiting discovery, and that there is a sufficient degree of uniform consistency in the working of the universe to permit successful argument by induction as a means to discover aspects of that order, despite the inevitably limited and particular character of the experience that motivates the belief.

This section has sought to set out considerations that present a reasoned defence of the realist interpretation of science. This philosophical conviction arises out of the actual experience of doing science, with its repeated feeling of satisfying discovery, rather than from a logical argument purporting to show that the world had to be open to our enquiry in this manner. The deep intelligibility of the universe is a fortunate fact, a wonderful gift that makes science possible. The deeper significance of this gift is a question to which we shall have to return in due course. Meanwhile, the kind of issues we have been discussing make it clear that scientific realism is something more subtle and more interesting than just naïve objectivity of the kind that an Enlightenment belief in access to clear and certain ideas might have encouraged one to expect. At the same time, acknowledging the subtlety of scientific belief should not drive us to embrace a post-modern account of science as social construction, as if its insights were the result of a largely unconscious and arbitrary choice by the invisible college of scientists, selected from a large portfolio of equally possible ways of thinking. Rather, the insights of science arise from and are controlled by our encounter with the way the world is, but in a complex and delicate manner that requires us to speak of scientific realism under the rubric of *critical realism*. The noun asserts the positive relationship of scientific knowledge to the way the world is, while the adjective acknowledges the subtle role that circularity and commitment play in its practice. Science yields well-motivated beliefs, but it does not deliver complete and absolute certainty about them. It is no stranger to belief in unseen realities—for example, quarks are thought to be 'confined' within the particles that they constitute, so that a single quark will never be observed in isolation. The existence of quarks must be defended by appeal to the intelligibility that they offer of more directly accessible phenomena (the properties of the particles that are made of quarks). In fact I believe that critical realism is a concept that is fundamental to the entire human quest for truth and understanding and that theology can defend its belief in the unseen reality of God by a similar appeal to the intelligibility that this offers of the general nature of the world and of great swathes of well-testified spiritual experience. A sophisticated twentieth-century approach of this kind can be found in the writing of Bernard Lonergan,[4] whose thought was shaped by the theological tradition stemming from Thomas Aquinas.

At this stage, a final point remains to be made. Discussion in later chapters will show that when one comes to very broad issues about the character of reality, such as the nature of time and the nature of causality, while our thinking is constrained by scientific knowledge it is not totally determined by it. We shall see that there remain judgements to be made which require acts of metaphysical decision. Scientists often eschew the idea of metaphysics and claim to have no need of it, but later in this chapter I shall seek to show the indispensability of metaphysical thinking for anyone wishing to attain an integrated world-view.

4. B. Lonergan, *Insight*, Longman, 1958.

THE NATURE OF THEOLOGY[5]

If science is human reflection on impersonal encounter with
the physical world, theology is reflection on transpersonal encounter with the sacred reality of God. It is immediately apparent that this is likely to be a much more difficult and subtle
task even than that pursued by science. We transcend the
physical world and can put it to empirical testing through the
contrivance of experiments. In science the initiative of discovery lies largely with the experimenter. God transcends finite humanity and is not open to experimental manipulation.
To suppose the contrary is to commit the sinful error of attempting magic. 'You shall not put the Lord your God to the
test' is just an inescapable condition of true encounter with
divine reality. In those acts of divine disclosure that theology
calls revelation, the initiative lies with God. Moreover, finite
minds will never be able to capture the Infinite adequately in
their logical nets. There is a tradition in theology, called apophaticism, which warns against the hubris of claiming exact
knowledge of deity. Yet belief in the existence of revelatory
divine self-disclosures means that this insight is not a counsel
of despair, but simply a caution about the degree of success that
theology can expect to attain. In consequence, the language
of theology will have to be the allusive and open language of
symbol rather than the precise language of mathematics that
is so effective in science. To a significant degree in theology,
prosaic clarity has to give way to something more like poetic
discourse. Thus the search for truthful understanding is more
difficult theologically than it is in science, but it is not impossible, as later discussions of the grounds for particular theo-

5. OW 3; SC 6; RR 4–5; SCB 3; FSU 1–2; TCS 1.

logical beliefs will seek to illustrate. Fundamentally, the two disciplines of enquiry should be thought of as cousins under the skin because of their shared truthful intent. Both operate under the rubric of critical realism, claiming the attainment of well-motivated beliefs, but not asserting the achievement of absolute certainty.[6] The religious recognition of this fact is expressed in the understanding that believers walk by faith and not by sight. Like Michael Polanyi in the case of science, the beliefs of religious people are sufficiently well-motivated for them to be able to commit themselves, despite knowing that in principle they might be mistaken. Religious faith does not demand irrational submission to some unquestionable authority, but it does involve rational commitment to well-motivated belief.

Having said this, we must also acknowledge some further significant differences between truth-seeking in science and in theology. Despite the role of personal skills and judgement in the practice of science, the investigator is able to adopt a detached attitude to the actual objects of his or her enquiry. Theology, like any form of personal encounter with reality, must take the risk of a more vulnerable kind of engagement. God is not to be met with simply in a spirit of intellectual curiosity, but with openness to the experience of awe and a demand for obedience. Religious knowledge is much more 'dangerous' than scientific knowledge, for it can imply consequences for the way we live our lives, requiring not only the assent of the intellect but also the assent of the will.

The impersonal dimension of science means that it is a linearly progressive discipline in which knowledge and under-

6. See the views of scientist-theologians summarised in S as T 2.

standing accumulate from generation to generation. Any physicist today understands much more about the universe than Isaac Newton ever did, simply by living three centuries later than that great genius. In religion, as in every other kind of personal encounter with reality, there is no presumption to be made of the superiority of the present over the past. Just as the individual creative work of Bach and Beethoven is an indispensable part of our present experience of music, so in theology the insights of great figures of the past—Augustine, Aquinas, Calvin and the rest—remain a necessary part of the contemporary conversation. There is no necessary implication of the superiority in every respect of twenty-first-century theological insight over that of earlier centuries, any more than there is of contemporary music over that of the past. Theological thinking has to be prepared to span the centuries in a way that is not paralleled in science.

Theology's anchorage in human encounter with the divine means that it is more sensitive to experiential context than is the case for science. Deeply personal experience will always be enabled and articulated within a specific cultural setting, which both offers opportunities of insight and imposes possible constraints of perspective. Part of the richness of theological thinking arises from its including within itself specific contextual theologies, grounded in the particular experiences of specific communities: feminist theology (based on the insights of women and often severely critical of what it perceives to be a male-dominated Church); liberation theology (drawing on the insights of the poor, especially in developing countries, and oriented to a demand for social justice); and so on. The present volume is an exercise in doing theology in the context of science.[7] Further theological complexity arises from a distinction, to be explored later, between natural theology (appealing to aspects of general experience) and revealed theology (appealing to particular persons and happenings, held to be specific occasions of divine self-disclosure). Scripture is to be understood as the record of these latter revelatory events—of course requiring interpretation for reception of the truths that they carry—and not as the delivery of propositional truths to be received as having non-negotiable verbal authority. Despite the diversity of its component parts, theology can nevertheless claim to be fundamentally a single integrated discipline, ultimately reflecting the unity of the God of whom it seeks to speak, just as science is essentially a unity, reflecting the unity of the natural world, despite the particular characters of its component disciplines of physics, biology and so on.

It is a central thesis of this book that both science and theology can lay claim to the achievement of a degree of truthful understanding that warrants their insights being described under the rubric of critical realism. Yet one must also acknowledge that the adjective 'critical' has the stronger force in the case of theology, because of the profound nature of its subject matter. The fact that theology is concerned with the acquisition of motivated belief, rather than the assertion of fideistic certainties, means that it is open to development and correction in its understanding. The history of Christian thinking from the New Testament to the great Councils of the fourth and fifth centuries, which articulated the doctrines of the Trinity and the two natures of Christ, shows just such

7. TCS.

a character in the process of the clarification of belief and the correction of heresy. In fact the challenges of those later deemed as heretical played an important role in provoking those later deemed orthodox to seek clearer understanding of the profound truths with which both groups were seeking to wrestle. The development of theological thinking has continued in the centuries that followed, down to the present day. Religious belief has not proved immune to the need for correction, even if the pace of change has sometimes been slow. It took the Christian community 18 centuries to recognise that the institution of slavery was inconsistent with human dignity, and rather longer to question whether a loving God would exact eternal punishment for finite sins.

The theological discussions in the chapters that follow will seek to illustrate and clarify the nature of truth-seeking activity on the part of theologians.

RATIONAL STRATEGY[8]

The essence of rationality is to seek to conform one's thought to the nature of what is being thought about. Science makes it clear that there is no single form that such rationality has to take. We think about the clear and orderly world of Newtonian physics in one way, but we have to think about the cloudy and fitful quantum world in a different fashion, in a manner quite counterintuitive to the expectations of everyday understanding. Different logics apply in these two domains. The everyday logic of Aristotle is based on the law of the excluded middle, requiring that there is no possibility intermediate between the two extremes of A and not-A. The billiard ball is

8. SC 6; RR 1–2; SCB 2; BG 2; S as T 2; TCS 2; QPT.

either here or it is not here. In the quantum world, however, we have seen that the superposition principle allows an infinite range of intermediate possibilities, formed of mixtures of the state in which the electron is 'here' (A) and the states in which it is 'elsewhere' (not-A). Consequently, in the quantum world a different, quantum logic has to apply. It is scarcely surprising then that theology also calls for its own form of rational discourse. Christian belief centres on the conviction that in Jesus Christ the truly human and the truly divine are both present. Here is a duality even more counterintuitive than the wave/particle duality of light. Of course, the strangeness of the latter does not explain or license the strangeness of the former, but there is at least encouragement to think as boldly as experience has been found to demand of us.

There is also no single epistemology. In science we can know the Newtonian world in all its clarity, but the quantum world has to be known in accordance with its Heisenbergian uncertainty, so that if we know where an electron is we cannot know its momentum (how it is moving), and if we know its momentum we cannot know where it is. The ways in which we know persons, and the way in which we know God, are different again from the ways in which we know the impersonal objects of science. While there may be analogies between ways of knowing persons and ways of knowing God, they are certainly not identical. True knowledge of God must be open to the experience of awe, the duty of worship and the divine demand for obedience.

I believe that science and theology both require the rational strategy that I have called 'bottom-up' thinking, seeking to move from experience to the attainment of well-motivated belief and understanding, rather than relying on a 'top-down' approach based on the hope that one has prior access to clear and certain general ideas from which one can then descend to the consideration of the particularities of experience. I do not assert that there is no place for top-down thinking, but I do believe that it must always be open to bottom-up evaluation and critique. The unexpected and surprising strangeness that science has so often encountered in its exploration of the physical world does not encourage reliance on the top-down approach. Allegedly 'clear and certain' ideas have often proved to be neither clear nor certain. In consequence, the natural question for a scientist to ask about any proposed belief is not 'Is it reasonable?', as if we felt we knew beforehand the shape that rationality had to take, but rather 'What makes you think that might be the case?' This form of enquiry is open to surprise and it does not seek to lay down beforehand the character of an acceptable account of reality. Yet it is also demanding, for the answer given will only be acceptable if motivating evidence is offered in its support. I think this strategy of bottom-up thinking is also to be followed in theology and later discussions will seek to give some examples of how this may be done. The very fact of the use of this approach in theology is the reason why I place it in the spectrum of the human search for truth achieved through *motivated* belief.

Bottom-up theological thinkers reject the claims of fideism to have access to indubitable knowledge of the divine, mysteriously conveyed in the form of infallible propositions that are endowed with unquestionable authority and immune from challenge or critique. The discourse of theology is not concerned with 'proofs' of God's existence that only the stupid could deny. In fact, even in science the concept of certain proof is seldom appropriate. Recall Polanyi's statement that scientifically he was able to commit himself to what he believed to be true, though he knew it might be false. The bottom-up thinker in science or theology lives by reasonable faith but not by certain sight. Even in mathematics a degree of commitment is called for, since Kurt Goedel has shown that axiomatised systems cannot establish their own consistency by means of internal argument.

In their explorations of reality both science and theology have recourse to the use of models in order to gain partial insight. A model is based on a picture of reality which reproduces certain features thought to be relevant for addressing a particular issue, without pretending that the model offers a fully ontologically adequate account of the nature of a complex reality. Models offer valuable partial insight, but not complete understanding. This means that one may often employ a variety of different models of the same entity, useful for different purposes, even if these models would be mutually incompatible if taken to be literally precise. For example, in nuclear physics, when concerned with nuclear fission it is helpful to picture the nucleus as a 'liquid drop', while discussion of the scattering of particles by a nucleus is better understood in terms of the picture of a 'cloudy crystal ball'. Of course, the nucleus is, in fact, neither of these things. Ultimate understanding requires the eventual replacement of a portfolio of incompatible models by a single integrated theory.

Theology also uses models, for example, pictures of God as righteous Judge and as loving Father, but the challenge of its intellectual task means theological theory-making is much more difficult and only a limited degree of success can reasonably be expected. Once again we must remind ourselves of the warning of apophatic theology that that Infinite Reality will never be fully captured by finite human thought. Acknowledgement of this is not to devalue theology, but to recognise its intrinsic character.

SCIENCE AND THEOLOGY[9]

I hold a passionate belief in the unity of knowledge. Therefore I believe that one must look beyond the insights achieved by the individual disciplines of enquiry, such as science and theology, to seek an integrated account of the whole of reality. Pursuing this desire leads to the consideration of further issues.

The first is to begin to seek an understanding of how science and theology relate to each other. Ian Barbour offered a taxonomy of possibilities that many have found helpful.[10] He outlined four possible stances, which he labelled conflict, independence, dialogue and integration. Conflict corresponds to the situation in which one or other discipline asserts the claim to be the only source of worthwhile truth and understanding. Either it is science that is said to answer all questions that are meaningful to ask and capable of being answered, or theology is said to be in possession of an exclusive key to knowledge that enables it to give authoritative answers even about issues such as the age of the Earth and the history of life. These claims are vociferously maintained today by fundamentalists of one conviction or the other, but both positions are, in fact, perverse. A good deal of the fairly widespread

9. BG 4; S as T 7; ST 1; TCS 5.
10. I. G. Barbour, *Religion in an Age of Science*, Harper Row, 1990, ch. 1.

belief in society today that science and religion are engaged in a battle to the death arises from the crude claims of what is supposed to be either a totally omnicompetent science or an infallibly omniscient religion. An honest science addresses only one set of questions (roughly How?—concerned with the processes of the physical world), while theology addresses another set (roughly Why?—concerned with the meaning, value and purpose present in what is happening). Neither side can claim to answer the other's questions, but we are perfectly familiar with the fact that both kinds of question are meaningful and necessary to ask. The kettle is boiling both because gas heats the water (the scientific explanation) *and* because I want to make a cup of tea (an explanation invoking purpose). We do not have to choose between these two accounts, for both are true. Without taking the two of them together, the event of the boiling kettle would only be partially understood. If we are truly to understand the rich, many-levelled world in which we live, we shall need the insights of both science and religion.

Recognising the different kinds of question that science and theology address has led some to take the stance that Barbour calls Independence. Science and theology are then said to be so distinct from each other that there is no possibility of interaction between them. Each goes its own way, freely exploring the two disjoint realms of insight to which they refer. But this is a highly implausible claim. It is true that How? and Why? are different questions, but the ways in which they are answered must be consonantly related to each other. Putting the kettle in the refrigerator is clearly incompatible with the claim to want to make a cup of tea! It is surely clear that science's discovery of evolutionary processes acting over vast spans of deep time has influenced the tone of theological discourse on the world as a divine creation, without at all having negated the possibility of that discourse.

Dialogue is the stance that recognises that there has to be consonance between the perspectives on reality offered by science and theology if both are indeed truth-seeking endeavours, and therefore there must be a mutually respectful interaction between the insights of the two. The ways in which they answer their separate questions must be congruent with each other. The resulting binocular vision onto reality may be expected to yield a view that is deeper and more comprehensive than either discipline could offer on its own. This is the premise on which the enquiry pursued in this book is based, presented as an expression of its author's belief in the ultimate unity of knowledge.

The stance of Integration seeks to carry this interaction further with the ambitious aim of attaining a fully unified synthesis of science and theology, merged into a single discipline. The danger in this project is that the synthesis will in fact be achieved by one discipline taking the dominating role, so that the other is simply assimilated to its partner's style of thinking. A better strategy is the even-handed quest for a theistic metaphysics, within whose wide embrace both science and theology can both find their proper place without prejudice to the status of their individual insights.

We have already noted that metaphysics is not a word that many scientists feel very happy with. It is not uncommon for the concept to be dismissed with the remark that the writer has no time or use for the notion of metaphysical thinking. In actual fact, it is impossible to think seriously without taking a metaphysical stance, since this simply means adopting a world-view. We think metaphysics as naturally and inevitably as we speak prose. The physical reductionist who claims that there is nothing but matter and energy, and no truth but the truth of science, is making a metaphysical statement as clearly as someone who looks at the world from a theistic perspective. The reductionists have not derived their belief from science alone. Everyone, implicitly or explicitly, has a metaphysics.

Scientism is the metaphysical belief that science tells us all that can be known or is worth knowing. It must clearly be distinguished from science itself which, owing to its intrinsic limitation to only a certain kind of encounter with reality, is far from being in a position to make such an overblown claim for its explanatory power. Science has bracketed out too much (meaning, purpose, beauty) from its consideration for it to be the universal source of understanding.

Every metaphysical scheme has to rest on a defining basis, which is not itself explained, but which is assumed as the foundation for the subsequent explanations that flow from it. In the tradition of Western thought, there are, broadly speaking, two choices for this foundational assumption. One takes as its assumed basis the brute fact of the properties of matter; the assumed basis of the other is the brute fact of the existence of a divine Agent or Creator. The first choice corresponds to materialism; the second choice corresponds to theism. Each choice has to defend itself by seeking to show that it provides the most economic, coherent, adequately comprehensive and intellectually satisfying understanding of the rich range of human experience of reality. In neither case can there be a claim to attain indubitable proof of the point of view adopted, but instead warrant must be sought by seeking to show that this metaphysical perspective affords access to the 'best explanation' of the nature of reality, a claim to be assessed in terms of the achievement of economy, naturalness of explanation and full adequacy to experience.

Theology practised in this metaphysical mode is often called philosophical theology, in contrast to a more narrowly defined reflection focused on religious experience and insight, which is called systematic theology. Science and systematic theology are both first-order disciplines, engaging with the specific dimensions of the human experience of encounter with reality that are their defining concerns, and seeking to respond to the questions that arise from these concerns. Philosophical theology is a second-order project, metaphysical in character, aiming to articulate a comprehensive world-view. In that role, it has to take seriously the insights of all the first-order disciplines, without pretending that it is in a position to exercise a right of correction over the conclusions that each has reached in its proper domain. The task of philosophical theology is to take these conclusions and incorporate them in an account that affords the widest and most profound context of truthful understanding, based on belief in the existence of God.

A positive evaluation of the interaction of science and theology will aim at exhibiting a consonant relationship between the two, expressed through a theistic metaphysics. The resulting view of reality will take a form significantly shaped by the content of the relevant theological component and I have suggested a taxonomy that reflects this fact.[11] Four broad approaches seem possible. Deism simply sees God as the Great Architect of the universe, the One who ordained its wonderful

11. ST 1.

order but, having set the worlds spinning, then simply left cosmic history to unfold. The stance labelled Theism allows some concept of continuing divine concern and interaction with creation, but it sits comparatively lightly to the specific insights of any particular religious tradition, such as Christianity with its belief in the resurrection of Christ. A Revisionary stance takes tradition seriously, but considers that its insights are likely to need radical modification in the light of modern knowledge. A Developmental stance acknowledges that theological discourse will be influenced by modern discoveries but believes that this can happen in a way that maintains significant continuity with the foundational insights of the past. The discussion that follows offers resources for evaluating these different approaches.

CHAPTER TWO

Some Lessons from History

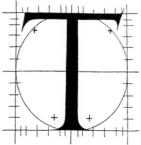HE bottom-up thinker will want not only to ground discussion of the interaction between science and theology on the character of these two disciplines, but will also seek to draw insight from the details of the way in which that interaction has actually taken place over the centuries. The story is a complex one, as John Hedley Brooke has made clear in his careful scholarly survey,[1] and it certainly does not lend itself to summary in terms of some single descriptive category, such as conflict or harmony. Fully adequate treatment is not possible in a short chapter, but it will be helpful to analyse two historical incidents that in the popular imagination have been accorded almost mythical significance, and also to consider an interesting proposal about the role of religious influence on the original development of modern science in Western Europe.

1. J. H. Brooke, *Science and Religion*, Cambridge University Press, 1991.

The first incident centres on Galileo. He is unquestionably one of the great founding figures of modern science, someone who pioneered the combination of mathematical analysis and experimental observation in the search for a deep understanding of the nature of the physical world. In 1609 he made a telescope for himself and turned it away from the Earth and onto the heavens. The resulting discoveries of mountains on the Moon, satellites encircling Jupiter, the phases of Venus, and a myriad of stars too faint to be seen with the naked eye, transformed human understanding of the universe. Galileo was confirmed in his belief in the heliocentric Copernican system, though he apparently submitted to the requirement of the Church not to teach or defend it in public. However, in 1632 he published his *Dialogue Concerning the Two Chief World Systems.* Formally the discussion did not adjudicate between Ptolemy and Copernicus, but the arguments presented in support of the latter's point of view were so strongly stated that there could be no doubt where Galileo believed the truth to lie. This brought him into immediate trouble with the Church authorities, who supposed that the Bible required the belief that the Earth was stationary. Galileo was called to appear before the Inquisition in Rome. He was never tortured, but he was required to recant his Copernican beliefs. A sentence of life imprisonment for his alleged errors was immediately commuted by Pope Urban VIII to one of house arrest.

This is certainly an unedifying story of prejudiced ecclesiastical response to a scientific discovery. It took the Roman Catholic Church a long time to put the matter right, with the formal ban on Copernicanism only being rescinded in 1820 and Galileo himself only being fully rehabilitated in modern times. Yet the issues involved were too complex for it to be possible to treat the incident simply as a stark confrontation between scientific light and religious darkness. Believers were to be found on both sides of the argument. Galileo himself was a religious man and he invoked in his defence the claim that the role of scripture is to tell us how to go to heaven, not how the heavens go. Part of the trouble was that the Church had invested too strongly in the philosophy of Aristotle and it had failed to heed the advice of St Augustine in the fourth century that if well-established secular knowledge seemed at odds with a current interpretation of scripture, then it would be wise to reconsider that interpretation. There were also personalities involved in the conflict. Galileo made brilliantly polemical use of the Italian language and the Pope was offended by the fact that the speaker on behalf of Ptolemy in the *Dialogue*, who was called Simplicio, bore a recognisable resemblance to himself. There were also genuine scientific difficulties. If the Earth was moving, why was there no observation of parallax among the fixed stars? It was not altogether easy at that time to believe that they were at the immense distances that this would have to imply. Galileo was completely mistaken in his explanation of the tides, although he regarded it as one of the pillars of his theory. Galileo's chief opponent, Cardinal Bellarmine, urged upon him that he should regard the Copernican system simply as a convenient way of 'saving the appearances', that is to say, just a way of doing the calculations without implying a belief that it described the way that things actually were. Ironically, such a non-realist account of science, seen just as instrumentally useful but not ontologically serious, would resurface in twentieth-century philosophy of science. In the previous chapter I gave reasons for rejecting it in favour of a realist interpretation.

With hindsight we can see clearly that in the Galileo case the Church made some bad mistakes, but these should serve as warnings of the need for future carefulness rather than discrediting the possibility of positive interaction between science and theology. Yet do we not see the same thing happening all over again in 1859, when Charles Darwin published *On the Origin of Species?*

Here again, the popular mythic account alleges a picture of stark confrontation and conflict, implying that, while Darwin's ideas received general welcome in the competent community of science, they were solidly opposed by religious people. This is simply historically false. There was a mixed reaction in both communities. There were significant doubts among many scientists and Sir Richard Owen, the greatest comparative anatomist of the day, never accepted Darwin's ideas. A big problem was the nature and origin of the small differences between successive generations on which natural selection had to be supposed to act. In fact, the answer to this difficulty was discovered only a few years later, by the Austrian monk Gregor Mendel. However, he published his genetic discoveries in an obscure central European journal and their true significance was not recognised until the twentieth century. On the religious side, there were many who welcomed Darwin's ideas. The Oxford theologian, Aubrey Moore, said that Darwin, in the guise of a foe, had done the work of a friend. Moore meant that in an evolving world God would be understood, not as a distant and detached Creator but as intimately working in the world through the unfolding of divinely ordained natural processes. Darwin's clergyman friend, the novelist Charles Kingsley, coined a phrase that succinctly sums up the illuminating theological way in which to think about the scientific fact of evolution. Kingsley said that no doubt God could have brought into being a ready-made world, but Darwin had shown us that the Creator had done something cleverer than that, in bringing into being a world so endowed with fertility that creatures could be allowed to explore and bring to birth its potentialities, in a process in which they 'made themselves'. At the very meeting of the British Association for the Advancement of Science at Oxford in 1860, at which the celebrated confrontation between 'Darwin's bulldog', Thomas Henry Huxley, and the Bishop of Oxford, Samuel Wilberforce, took place, the young Frederick Temple, later to become Archbishop of Canterbury, preached a sermon in which he welcomed Darwin's ideas.

As the nineteenth century progressed, the reception of evolution in Christian circles became increasingly widespread. The highly conservative President of Princeton Theological seminary, B. B. Warfield, was to say that he detected no fundamental contradiction between evolution and Genesis. Darwin himself, though not a Christian believer, had affirmed that he thought it was perfectly possible for someone to be both a theist and an evolutionist. It was only in the first quarter of the twentieth century that rejection of Darwinian ideas began to grow in fundamentalist Christian circles.

So much for these two alleged critical partings of the ways between religion and science. A counterbalancing positive historical assessment of the interaction between science and theology can draw support from reflection on the origins of modern science itself. It developed in Western Europe from the late Middle Ages onward, bursting into first full flower in the seventeenth century with the discoveries of Galileo, Kepler and Newton. It is interesting to ask why this happened then and there and not among the ancient Greeks, a people displaying great intellectual power and curiosity, or among the medieval Chinese, whose civilisation was for so long in many respects in advance of that of Europe. Historically counterfactual questions of this kind (What would have happened if this or that had been different?) cannot receive indisputable answers, but there is a respectable case to be made that the Judaeo-Christian understanding of the world as God's creation was a significant influence aiding the development of science.[2] To believe that the rational God was the Creator of the world encouraged the belief that its structure would display a deep order which would be accessible to human reason if humans are indeed beings made in the image of God. Of course, the Greeks believed that the universe was rationally ordered, but they thought that its creator, the demiurge, had to follow an eternal pattern already existing in the noetic world of Ideas. If that were true, then the most direct way of discovering that order would be through pure contemplative thought, rather than messy empirical observation. Jews and Christians, in contrast, believed that God was completely free to endow creation with whatever order was pleasing to the divine will. In consequence of the Creator's freedom, one had to look and see what was the order that God had actually chosen to bring about. One could not argue on abstract grounds how falling bodies ought to behave, but one must experiment, as Galileo did, to find out how they actually

2. S. L. Jaki, *Science and Creation*, Scottish Academic Press, 1986.

fell. The doctrine of creation also implied that nature itself was not divine and so there would be no impiety in its manipulation and empirical probing. Yet nature possessed a value in itself precisely because it was God's creation, and so its exploration was a worthy project. Perhaps this latter point was one that did not come readily to the Chinese, with their attention so greatly focused on the ordering of civil society.

It is certainly the case that most of the founding figures of modern science were people for whom religious belief was important, even if some had problems with the Church authorities (Galileo) or with Christian orthodoxy (Newton). Sir Isaac himself saw the marvellous order of the solar system as a clear sign of the designing Intelligence who was its Creator, and he believed that if occasional angelic intervention was required to maintain its stability this would be provided. In the seventeenth century it was a common thought to say that God had written two books, the book of scripture and the book of nature. Both were to be read, and if this was done correctly there could be no contradiction between them, since they had the same Author.

In the eighteenth century, however, Newton's successors became so bewitched by the deterministic nature of his equations that they began to believe in a closed world of mere mechanism, even writing books with titles such as *Man the Machine* (de la Mettrie). For many this stance led to the repudiation of theism, for God had come to be reckoned an unnecessary hypothesis. We shall have to return to this issue later.

CHAPTER THREE

Insights from Science

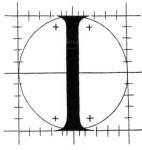F science and theology are colleagues in the common quest for truth, then they will have cousinly gifts to offer to each other. What science can give to theology is to tell it what the history of the universe has been like, including the history of life on Earth, and what is the nature of the physical world. These are insightful gifts of great value, which theology should be glad to receive. It is sad to see the spectacle of some Christian believers turning their backs on science. Those who seek to serve the God of truth should welcome truth from whatever source it may come. Science is certainly not able to tell us all the truth, but it does have insights to offer that deserve to be treated with the greatest seriousness and to be received with gratitude.[1] Looking into six areas of scientific insight will provide some indication of the value of what science has to offer.

1. OW 4; SC 3; RR 3, 7; ST 3; ER 2.

In this chapter I shall be content to outline the nature of these scientific understandings, leaving till later in the book the task of drawing out what consequences for theology might flow from them.

CAUSALITY

After the great discoveries of Sir Isaac Newton, it seemed to many people in the eighteenth century that science described a mechanical world of tightly deterministic causality. There was always something strange about this idea, for it is a basic human intuition that we are not mere automata. Not only would such a belief negate the widespread conviction that human beings exercise a significant degree of freedom in choosing how to act, but it would also threaten to subvert belief in human rationality. If we were simply computers made of meat, what would validate the program running on our hardware? No doubt the evolutionary need for survival could be appealed to as having shaped our brains so as to be able to cope with everyday experience, but to suppose that this was sufficient to lead to the human capacity for highly abstract thought in mathematics and philosophy would be to make a rash and implausible conjecture, since these capacities confer no obvious extra ability for survival in a purely physical environment. The richness of human life only begins to make sense if mental experience is more than epiphenomenal froth on the surface of a purely materialist reality. If we were simply machines, human discourse could be no more than automatic mouthing, with no true exchange of rational argument taking place.

The discovery of field theories in the nineteenth century did not significantly alter the arid prospect that physics seemed to offer. It is true that the world was no longer seen as populated simply by billiard-ball atoms colliding in the void, but the equations of classical field theory are as strictly mathematically deterministic as are the equations of Newtonian particle dynamics. However, the twentieth century was to see the death of a merely mechanical account of the physical world.

This came about through the discovery of intrinsic unpredictabilities present in nature. The qualifier 'intrinsic' is important here. These are not unpredictabilities that could be removed by measuring a bit more accurately or calculating a bit more precisely, for they are properties of nature itself. It was the discovery of quantum theory that first brought this phenomenon to light. The world of subatomic physics proved to be cloudy and fitful in its character, in complete contrast to the clear and reliable everyday world of Newtonian mechanics. Heisenberg's uncertainty principle stated that empirical knowledge of the quantum world was much more restricted than that assumed to be accessible in the realm of classical physics. Newtonian thinking had supposed that one could know exactly both the positions and the momenta of all particles involved. At the level at which quantum effects are important, Heisenberg showed that trying experimentally to obtain exact knowledge of position would induce unavoidable total ignorance of momentum, and vice versa. You could know where an electron was, or how it was moving, but you could not simultaneously have exact knowledge of both. While the Schroedinger equation, the basic equation of quantum physics, is a perfectly deterministic differential equation, it does not relate to directly observable particle properties, but it determines the wavefunction from which only the probability of observing some particular physical property can be calculated. Most of the information that it carries relates to potentiality rather than to definitely realised actuality.

Whatever the nature of the physical world might actually be, it had been discovered to be something more subtle and more interesting than just a domain of cosmic clockwork. Quantum physics implied the illusory character of the dream that Laplace had entertained, of a calculating demon who, having obtained exact and total knowledge of the present, would then be able fully to predict the future and fully to retrodict the past. The critical issue is whether these discoveries mean that physical reality is not only subtle but also supple and open to the future.

There is no question that quantum physics has turned out to be probabilistic. The quantum physicist can calculate the chance of a radioactive nucleus decaying within the next hour, but it is not possible to determine whether a specific nucleus will actually undergo decay within that period. However, probabilities can arise for two quite different reasons. One is simply ignorance of all the factors which, if they were known, would in fact be sufficient to determine what will happen. The fall of a die is a paradigm example. Maybe all those radioactive nuclei have, so to speak, hidden internal clocks whose settings determine exactly when each nucleus will decay, but which the experimentalist cannot get to read. The other possible source of probability is the presence of a radical indeterminism in the processes of nature, at least as far as science is able to describe them. The first of these possibilities corresponds simply to epistemic ignorance, but the second possibility corresponds to the ontological reality of an irreducible degree of openness present in natural process. In the quantum case, the question is whether Heisenberg's uncertainty principle is simply a principle of unavoidable ignorance or is it a principle of true indeterminism?

The pioneers of quantum theory, guided by Niels Bohr and adopting his celebrated Copenhagen interpretation, almost all opted for the second alternative. Succeeding generations of quantum physicists have mostly followed in their train. Paul Dirac identified the cloudy fitfulness of quantum physics as originating in the superposition principle, which permits such counterintuitive mixtures as the existence of states in which an electron is both 'here' and 'there', so that in such states it has no definite position.

However, in the 1950s, David Bohm showed that there is an alternative formulation of quantum theory which yields exactly the same experimental consequences as those of the conventional Copenhagen interpretation, yet it is fully deterministic in its basic structure.[2] Bohm achieved this remarkable feat by divorcing wave and particle, which Bohr had proclaimed to be indissolubly united as complementary aspects of individual quantum entities. In Bohm's theory there are particles which are as unproblematically objective and deterministic in their behaviour as Sir Isaac Newton himself might have wished them to be. However, there is also a hidden wave, encoding information about the whole environment. It is not itself directly observable, but it influences in a subtle and highly sensitive manner the motions of the particles in just such a way as to induce the experimentally observed probabilistic effects. For Bohm, the uncertainty principle is a prin-

2. D. Bohm and B. J. Hiley, *The Undivided Universe*, Routledge, 1993.

ciple of unavoidable ignorance of the infinitesimally fine detail of his deterministic world.

The empirical equivalence of these two interpretations of quantum theory means that the question of which one is to be preferred is not an issue that can be settled by physics alone. The choice between Bohr and Bohm is not a scientific matter, but it requires an act of metaphysical decision. Almost all physicists follow Bohr rather than Bohm because the latter's ideas, though very clever and instructive, have about them an air of being too clever by half. A good deal of the argument centres on technical issues, such as the initial conditions of the universe, the relation of Bohm's theory to special relativity, and the special role of position in its formulation, but something of the flavour of the reservations about Bohm's approach can be conveyed by asking what is the wave equation that Bohm's wave has to satisfy? It is perhaps not altogether surprising to learn that the answer turns out to be the Schroedinger equation. It is this choice that ensures empirical equivalence with Bohr. In conventional quantum theory, persuasive arguments can be presented in support of making Schroedinger's original choice, but in Bohm's theory the equation is simply borrowed ad hoc from Schroedinger himself. The resulting air of contrivance is a consideration that is held by many physicists to tell strongly against the metaphysical acceptability of Bohm's idea.

Bohr's Copenhagen interpretation of quantum theory corresponds to a world of becoming, open to its future. For example, suppose that before a measurement an electron was in a superposition of the states of being 'here' and being 'there'. After the measurement the electron's state is different, corresponding to the definite position actually observed. Physical reality has changed in a manner that could not be predicted beforehand.

In the 1960s there came the unexpected recognition of the existence of a further source of intrinsic unpredictability in physics. It turned out that even classical mechanics was not as tamely predictable as its deterministic equations had seemed to imply. (In fact this had been realised much earlier by the distinguished French mathematician, Henri Poincaré, but the significance of his discovery was not properly appreciated at the time.) This further source of intrinsic unpredictability came to be called 'chaos theory'.[3] These new unpredictabilities arise from the fact that certain systems can possess such an exquisite degree of sensitivity to their initial conditions that effectively their future behaviour is beyond calculation. Even an infinitesimal change in those initial circumstances would completely alter the subsequent behaviour. It is perhaps not surprising to learn that Ed Lorenz was led to this act of rediscovery by studying a simple non-linear equation that described aspects of the behaviour of weather systems. This fact led to the serious scientific joke of the 'butterfly effect'. The Earth's weather system might be in a state of such extreme sensitivity that a butterfly in the African jungle, stirring the air with its wings today, could create a disturbance that would grow and grow until it resulted in a storm over Europe in three or four weeks time! However, the name 'chaos theory' turned out to have been ill-chosen, since analysis shows that the behaviour of chaotic systems is not totally random, but it is constrained to lie within a wide range of possibilities, called a 'strange attractor'. Just as in the case of quantum physics,

3. See, J. Gleick, *Chaos*, Heinemann, 1988.

so in chaotic dynamics there arises the metaphysical question of whether this intrinsic unpredictability is to be considered as epistemic or ontological in its character. Of course, if the deterministic equations of Newtonian physics, from which the idea of chaos was mathematically derived, were physically exact, then what would be involved would simply be an unpredictability induced by epistemic ignorance of the finest details of the initial conditions. Yet we know that these Newtonian equations are only approximately true, corresponding to physical situations in which the entities involved are large and clearly separable from each other. Classical physics only offers an approximation to physical reality in certain circumstances. There is, therefore the possibility of an alternative, ontological interpretation of the fact of unpredictablity. This interpretation need not be taken to imply that future behaviour is just some sort of random lottery, but rather that there is room for the operation of novel additional causal principles bringing it about, beyond those described by conventional physics. This is an option to which we shall return in the next section.

For the moment we can note two important general lessons that can be drawn from this discussion. The first is that, while physics certainly constrains thinking about the nature of causality, it is insufficient of itself to determine what its character should be. As in the case of the decision between Bohr's indeterminism and Bohm's determinism in quantum physics, the selection of any theory of causality has to be made partly on the basis of satisfying metaphysical criteria, among which one could quite properly include theological or anthropological considerations in appropriate circumstances. The second lesson is to note that, because of this need for metaphysical decision, an honest science cannot claim to have established the causal closure of the world on its own restricted terms.

Finally, we must consider other aspects of science's exploration of the physical world which cast further doubt on any claim that it might make to be in a position to give a complete and universal account of the nature of causality. In specific regimes, such as the subatomic quantum world or the everyday world of Newtonian physics, science can offer many valuable insights. Yet there are serious unresolved problems in understanding how these different regimes relate to each other. For instance, we do not fully understand how the reliable macro-world of every day emerges from the fitful microworld of quantum physics. An unsolved problem in the interpretation of quantum theory is the measurement problem: what is the process that brings about a definite answer on each actual occasion of measuring a property of a quantum entity, such as the position of an electron or the momentum of a photon? The theory enables the physicist to calculate with stunning accuracy the relative probabilities of obtaining different experimental answers, but there is no wholly satisfactory and universally agreed understanding of how it is that a particular result is obtained on a particular occasion. What causes the electron to be found 'here' this time, rather than 'there'? Perhaps the greatest paradox of quantum theory is that, more than eighty years after the foundational discoveries, while we know how to do the sums we still do not understand all that is going on.

In fact, physics' account of the physical world is distinctly patchy. A simple illustration of the difficulty is provided by attempts to formulate 'quantum chaology', that is to say, to synthesise quantum theory and chaos theory. The extreme sensitivity of chaotic systems means that their future behaviour soon comes to appear to depend upon fine detail lying below the limit of the uncertainty principle. At first sight this might seem to offer an ideal opportunity to promote the effect of the widely believed openness of quantum processes so as to induce a similar openness at the level of macroscopic phenomena. The idea does not work, however, because the two theories are mutually incompatible. Quantum theory has a definite scale (there is a meaning to 'large' and 'small', expressed in terms of Planck's fundamental constant), but chaos theory is fractal in character and fractals are scale-free, looking essentially the same on whatever scale they are sampled. The two theories just do not fit consistently together. Problems of this kind indicate the need for humble realism about what science can actually tell us about the character of causal process.

RELATIONALITY AND HOLISM

Twentieth-century physics saw not only the death of mere mechanism but also the replacement of pure atomism by an increasingly relational conception of physical reality.[4] The most striking example of this tendency was the discovery of quantum entanglement.[5] Once two quantum entities, such as two photons, have interacted with each other, they can be in a state such that acting on one will produce an instant effect upon the other, however far they may have become spatially separated after their interaction. It is important to recog-

4. ST 3.
5. See, J. C. Polkinghorne, *Quantum Theory: A Very Short Introduction*, Oxford University Press, 2002, ch. 5.

nise that this counterintuitive connection (a 'togetherness-in-separation', or non-locality as the physicists say) is truly ontological in character and not just epistemological. It induces actual change. If there are two balls in an urn, one white and one black, and two people take them out in their closed fists without looking at them and, after parting, one subsequently opens his hand to find the white ball, he immediately knows that the other has the black ball. There is nothing at all surprising in this. What was always the case has simply become known. Quantum entanglement, however, involves genuine change brought about through distant causal connection. It is as if one person found a red ball, then the other would have had to find a blue ball, but if the first person's ball had turned out to be green, then the other's ball would have had to be yellow. The two entangled quantum entities have effectively become a single system, since they are never truly separated from immediate mutual influence. One might wonder whether this apparent action-at-a-distance does not violate special relativity. However, what the latter forbids is the faster-than-light transfer of *information* (such as might enable the synchronisation of clocks). Careful analysis shows that quantum entanglement does not permit this to happen. Each observer sees only a random result for measurement on a single entity and each is unaware of the correlations between the two sets of results which might have conveyed information between them. It is as if two singers were each singing a series of random notes. Only someone able to hear them both would know that the pairs of notes were strictly in tune.

Albert Einstein, with two young collaborators, had noted in the 1930s that quantum theory implied the phenomenon of entanglement, but he felt that it was too 'spooky' to be true in nature and so he supposed that the deduction simply showed the unsatisfactory character of conventional quantum physics (a theory which he had always disliked). It was only in the 1980s that clever experiments showed that entanglement is indeed a property of nature.

Earlier, Einstein's discovery of general relativity had shown that space, time and matter are intimately interrelated, as matter curves spacetime and that curvature in turn influences the motion of matter, yielding the modern understanding of the nature of gravity. The extreme sensitivity of chaotic systems to the slightest disturbance emanating from their environment shows that they can never properly be treated as isolated systems. One could write a history of modern physics as being variations on the theme of an increasing realisation that 'reality is relational'. It has turned out that even the subatomic world cannot properly be understood atomistically.

Another important development in the thinking of twentieth-century physics has arisen from the study of moderately complex systems, treated as wholes and not decomposed into their constituent elements. Of course, much has been learnt from pursuing the methodologically reductionist strategy of thinking in terms of the exchange of energy between constituents, but it has turned out that striking phenomena exist, involving the spontaneous generation of astonishing degrees of order, which can only be observed holistically and which are quite unforeseeable in terms of a constituent analysis. Many cases of this phenomenon are to be found in the behaviour of what are called dissipative systems, held far from thermal equilibrium by the exchange of energy and entropy (disorder) with their environment.[6] The continual export of disorder enables them to generate and maintain a high degree of internal order. A very simple example is provided by Bénard convection. Fluid is confined between two horizontal plates, the lower of which is maintained at a higher temperature than the upper. In certain well-defined circumstances, heat transfer between the two plates is carried by convective motion which is found to take place within an orderly array of hexagonal convection columns. This phenomenon involves the spontaneously correlated motion of trillions upon trillions of molecules of the fluid. Immensely more complicated examples are provided by living beings, all of which, from the point of view of physics, are dissipative systems.

A similar spontaneous generation of order has also been observed in computerised models of logical networks, which have been studied particularly by Stuart Kauffman.[7] One of these models (technically, a Boolean net of connectivity 2) can be pictured in physical terms as corresponding to an array of light bulbs. Each bulb has two possible states, 'on' and 'off'. The array is started off in a random pattern of illumination and it then evolves in steps according to some simple rules (which express the logical structure of the net). Each bulb has two correlates elsewhere in the array and its state at the next step is determined, according to the rules, by the present states of its correlates. One might have expected that usually nothing very interesting would happen and that the array would just flicker away randomly for as long as it was allowed to do so. In actual fact this is not the case, for the array soon

6. I. Prigogine and I. Stengers, *Order out of Chaos*, Heinemann, 1984.

7. S. Kaufmann, *At Home in the Universe*, Oxford University Press, 1995.

settles down to circulating through a very short sequence of patterns of illumination. This phenomenon corresponds to the spontaneous generation of an altogether astonishing degree of order. If there are 10,000 bulbs in the array, this limit cycle is found to consist of about 300 states, despite there being about 10^{3000} different states of illumination in which the array might be found.

In the case of these logical models, their deterministic nature implies that, for them, the observed holistic order must be wholly generated bottom-up from the relationships between the logical elements involved, even if its description is best expressed holistically. For physical dissipative systems, however, the presence of intrinsic unpredictabilities, capable of metaphysical interpretation as signs of a causal openness to the future, allows the possibility of conjecturing that in this case there is a genuine top-down influence acting from the whole upon the parts. This would mean that, in addition to the conventional physical picture of the exchange of energy between constituents, there are holistic constraints on the pattern of total energy flow within the system as a whole. The laws governing constituent interactions would then need to be complemented by holistic laws of nature expressing these top-down effects. As the saying goes, one would have to recognise that 'More is different', the nature of the sum exceeds the addition of the natures of the parts. The specification of these dynamical patterns of holistic behaviour can be called 'information'.

At present no general theory is known covering the emergent phenomena manifested by complex systems. Study is currently at the 'natural history' stage, with many intriguing particular examples known, but without an overall integrating theoretical point of view. Yet it is entirely reasonable to expect the discovery of such a general theory in the course of the twenty-first century, a development which can be expected to place information alongside energy as a fundamental category for physical thinking. The conjectured principle of top-down causality can properly be called 'active information'.[8] It would be reasonable to entertain the metaphysical hope that this development offers the prospect of offering some modest help towards beginning to understand the willed acts of agents. When a person raises his arm, there is of course a bottom-up story of currents in nerves inducing muscular contraction, but there is also a holistic story of the whole person deciding to enact that movement.

COSMOLOGY

The basic cosmological picture of the history of the observable universe, understood as a process of expansion from a singular point of origin (the big bang), is rightly part of every informed person's understanding today. There are excellent grounds for accepting this outline account, even if there remain points of detail over which scientific arguments continue. Our understanding of cosmic history has been greatly consolidated by a number of important discoveries made in the last 50 years.

Great advances have been made in the accuracy of the observations, resulting in a present estimate of the age of the universe at the rather precise figure of 13.7 billion years. The very early universe was highly energetic and for a while transformations took place with breathtaking rapidity. One of the

8. BG 3.

most important of these is believed to have been an era of inflation which set in when the universe was only about 10^{-35} seconds old. This process was one of extremely rapid expansion due to what the physicists would call a cosmic phase change, a kind of 'boiling of space'. There is theoretical motivation for belief in the possibility of inflation, even if there is still considerable argument about the details, including the mechanism by which the period of expansion was brought to a quick and orderly end. Powerful observational support for the idea comes from two otherwise puzzling aspects of the nature of our universe. One is that, averaged over distance, it is strongly isotropic, looking very much the same in whatever direction one might look. Without inflation, the different remote regions surveyed in the different directions would never have been close enough to each other to have interacted in a way that could have produced such uniformity. In the inflationary picture, however, initially they would have been close enough to be mutually interacting and in equilibrium, before being blown far apart. The second remarkable property is that the universe is almost flat on the largest scale, with expansive and contractive effects so nearly balanced that the resulting curvature of space implied by general relativity is very small indeed. An inflationary period in the very early universe would have had a smoothing out effect that would have led to just such flatness, whatever might have been the case before inflation acted.

After inflation, the universe continued to expand at a normal rate, at the same time cooling in the process. For about the first three minutes of cosmic history, the world was still hot enough for nuclear reactions to be taking place universe-wide.[9] It was, so to speak, a kind of cosmic hydrogen bomb. But then universe-wide nuclear reactions ceased because of the cooling induced by expansion, and the gross nuclear structure of the cosmos was frozen out in the state that it had then attained. Because the early universe was very simple, it had led only to simple consequences—in the case of nuclear matter, essentially the formation of the two simplest elements, hydrogen and helium. Theory predicts that they should have been formed in the ratio of 3 to 1 and this is indeed observed to be the case, a significant predictive triumph for big bang cosmology.

For several hundred thousand more years the universe was still hot enough for radiation to be sufficiently energetic to break up and stop the formation of stable atoms. In this era the universe was a kind of cosmic plasma, but when it had cooled sufficiently this period came to an end, the world-wide interaction of matter and radiation ceased, and the universe became transparent. The decoupled radiation then simply continued to cool with continuing cosmic expansion. This sea of radiation fills the entire universe. It was first observed in 1965 as the universal cosmic background radiation (CBR), now at a temperature of three degrees above absolute zero. The CBR is a kind of fossil record of the state of the universe when it was about half a million years old. It is almost uniform, but there are small variations of energy density at the level of one part in a hundred thousand. These inhomogeneities are extremely significant, for they were the seeds from which the subsequent grainy structure of galaxies and stars was able to grow. If the universe had been strictly uni-

9. See, S. Weinberg, *The First Three Minutes*, A. Deutsch, 1977.

form, it would have remained so indefinitely but, because at places where there was a little excess matter/energy there was a little stronger gravitational attraction, the inhomogeneities induced a snow-balling effect, eventually producing condensation into a series of separated structures. As a result, the universe became lumpy with stars and galaxies. As the stars formed and began to contract under their own gravity, they heated up and nuclear reactions began again in their interiors. These nuclear reactions not only kept the stars shining for billions of years, but they also generated new elements beyond hydrogen and helium. This process continued up to iron, the most stable of the nuclear species, thereby making many of the heavier elements that would eventually be necessary for the possibility of carbon-based life. Some stars at the end of their lives exploded as supernovae, not only scattering the elements they had made into the environment where they could become part of the next generation of stars and planets but also in the process creating the elements beyond iron, some of which are also necessary for life. We shall pay some more attention later to this delicate and beautiful process of nucleogenesis, whose unravelling was one of the great triumphs of astrophysics in the second half of the twentieth century.

Meanwhile we must turn back to consider the extremely early universe. It was initially so small that quantum effects and gravitational effects (general relativity) were equally important. The relevant epoch is indeed extremely early, before the Planck time of 10^{-43} seconds, but nevertheless it is a significant era in cosmic history. However, understanding it is problematic, for it is a remarkable and frustrating fact that quantum theory and general relativity, the two great discoveries of twentieth-century physics, are still imperfectly reconciled with each other. At present we are faced with a number of competing and speculative proposals about how to deal with this problem. Many interesting suggestions have been made, but their conjectural character means that all need to be received with a degree of cautious reserve. The currently most popular proposal for trying to reconcile quantum theory and general relativity is string theory.[10] It is based on the assumption of the existence of very tiny one-dimensional constituents ('strings' with a scale of 10^{-33} cm) vibrating in a spacetime of 10 or 11 dimensions, with some of these dimensions made inaccessible by being 'compactified' (rolled up) to yield the apparently four-dimensional spacetime of our direct experience. A great number of very clever ideas have gone into the formulation of string theory, but it is important to keep in mind its highly speculative character. The string theorists are claiming to be able to tell us about physical processes on a scale that is 16 orders of magnitude (powers of 10) smaller than anything with which we have had direct experimental contact. A leap over 16 orders of magnitude would take one from a town the size of Cambridge to something smaller than an atom. The lessons of history are not altogether encouraging to such bold ambition. Nature has so often manifested wholly unforeseen properties when investigated in new regimes previously unexplored.

Much of the interest in attempts to formulate an account of the extremely early universe while it was controlled by quantum gravity arises from conjectures that the origin of the big bang itself might lie in a quantum fluctuation taking place in a trans-cosmic ur-vacuum. This is an idea that re-

10. See, B. Greene, *The Elegant Universe*, Jonathan Cape, 1999.

quires some explanation. In quantum physics, the vacuum, defined as the state of lowest energy, is not simply an inactive void but a humming sea of transient fluctuations. The fundamental reason for this lies in the Heisenberg uncertainty principle. The essential point at issue can best be illustrated by thinking about a very simple physical system, the quantum 'pendulum'. In classical physics, the lowest energy state of a pendulum is that in which the bob is at the bottom and at rest. For a quantum pendulum this would imply that one knew both where the bob is (at the bottom) and how it is moving (at rest) which, of course, Heisenberg will not permit. Consequently a quantum pendulum, even in its lowest energy state, is subject to an irreducible degree of quivering (almost at rest, almost at the bottom), which the physicists call quantum fluctuations and which generates what they call zero-point energy. Analogous effects operate in the quantum fields that constitute the basic fabric of physical reality. The cosmological conjecture is that one of these fluctuations in the ur-vacuum was caught and expanded by inflation before it could disappear again, generating the big bang and our consequent universe. Ideas of this kind, obviously formulated with much more sophistication than this qualitative account can convey, have been hailed by some as amounting to the scientific achievement of something previously thought to have been a divine monopoly, creation out of nothing. However, a moment's thought will show that this claim is based on a gross abuse of language. An intensely fluctuating quantum vacuum is certainly not *nihil*. Even if the speculation were correct, the universe is not a 'free lunch', as people sometimes assert, for the cost of the cosmic meal is quantum theory itself and the quantum fields that are fluctuating.

A great deal has certainly been learnt about cosmic structure and cosmic history, yet a very great deal still remains
unknown. In recent years cosmologists have come to realise
that ordinary matter, of the kind well-known to physicists
and visible in stars and galaxies, is less than 5 per cent of the
matter/energy content of the cosmos. The observed stability
of rotating galaxies and clusters of galaxies implies that they
must be held together by a much stronger gravitational attraction than would be provided by their visible components
alone. In other words, they must also contain 'dark matter',
not interacting in the way that makes ordinary matter visible,
but exerting the same kind of gravitational force. There are
speculations that this dark matter, which makes up about 25
per cent of the universe, is composed of WIMPS (weakly
interacting massive particles) and that these are the so-called
supersymmetric particles that are postulated in most attempts
at formulating grand unified theories of the laws of nature,
but which have not so far been observed directly in any laboratory. The assumption being made is that all the observed
particles have supersymmetric 'partners', not yet discovered,
whose presence would endow the enlarged portfolio of particles with certain properties (supersymmetry) with desirable
consequences, such as making the construction of consistent
unified theories easier to attain.

The remaining 70 per cent of the matter/energy of the
universe is believed to be in the form of 'dark energy'. Observations have shown that the expansion of the universe, instead
of slowing down as a result of gravitational braking as was expected to be the case, is actually accelerating. The hypothesis
is that this acceleration is the effect of dark energy, an energy
associated with space itself. Einstein had recognised this as

a possibility when he discovered general relativity. However, later he came to regret the move, which he had introduced in a mistaken attempt to find equations that would describe a static universe, of the kind which at that time he believed would necessarily be the case. The zero-point energy of quantum vacuum fluctuations would generate just such an energy of space, but simple estimates of its magnitude based on this idea yield a value 10^{120} times greater than that actually observed! Therefore, if this concept of the origin of dark energy is correct, absolutely astonishing cancellations between the contributions from the different fluctuating fields must take place. Supersymmetry could be helpful in inducing at least some of these. We should be grateful for these cancellations, whatever their origin, since a cosmological constant (as the magnitude of dark energy per unit volume of space is called) which was even a little larger than that which is observed would have blown the universe apart too rapidly for any significant structures to have been able to form.

The small value of the cosmological constant is a particular example of how very precisely 'fine-tuned' the given magnitudes of the forces of nature have to be in a universe if it is to be capable of sustaining so interesting a development as the generation of carbon-based life. Many other examples are known of the highly specific character that must be possessed by a life-generating universe. They have been collected together under the rubric of the Anthropic Principle,[11] a not altogether happy choice of terminology since the conditions to which it refers are those that are necessary to allow the

11. J. Barrow and F. Tipler, *The Anthropic Cosmological Principle*, Oxford University Press, 1986; J. Leslie, *Universes*, Routledge, 1989; R. D. Holder, *God, the Multiverse, and Everything*, Ashgate, 2004.

emergence of any form of carbon-based life and, of course, they are not at all related to the specificity of homo sapiens. Two further examples of anthropic constraints can serve to illustrate the general kind of considerations involved.

Every atom of carbon in every living being was once inside a star, since the interior nuclear furnaces of the stars are the only places in the universe where this element can be made. The process of stellar nucleogenesis is both beautiful and delicate. Its first step involves the successive bonding of helium nuclei to form first beryllium and then carbon. However, the beryllium nucleus involved is extremely unstable and it would normally be expected to disappear too quickly to allow the attachment of a third helium nucleus to take place to produce carbon. This process is only possible because in our universe there is a very large enhancement effect (a resonance) in carbon occurring at just the right energy to enable the formation to occur with greatly increased rapidity. Only a small change in the intrinsic strengths of the nuclear forces would either have displaced the resonance to the wrong energy or eliminated it altogether, thus rendering the production of carbon impossible and the associated universe necessarily lifeless.

The second example of anthropic constraint relates to the size of the observable universe. Our local galaxy, the Milky Way, contains about 10^{11} stars, and there are about 10^{11} galaxies in the observable universe. We live in a world that is almost unimaginably vast. We do not know if there are life forms elsewhere within it, although it is very likely that there are many suitable planetary sites where this might be able to happen. The problem is that our ignorance of the process by which life first began on Earth means that we simply cannot estimate how likely it is to happen in another similarly propitious environment. In fact, very distinguished experts offer us widely differing opinions, ranging from 'almost impossible' to 'virtually certain'. Nevertheless, even if Earth is the only life-bearing planet in the universe, all those trillions of stars would be anthropically necessary, since only a universe at least as big as ours could last the fourteen billions years or so which is the natural time span required for the evolution of beings of our complexity. All scientists agree about the existence of anthropic fine-tuning in our universe, but what metascientific significance might thereby be implied is a matter of argument. This is an issue to which we shall return in the next chapter.

Cosmologists not only look back into the past but also peer into the future, at least to the extent of discerning its expected general shape. The prospect is not altogether encouraging. In about a billion years the Sun will begin the long process of turning into a red giant as it gradually uses up all its hydrogen fuel, a transformation that will burn up any life that might still be remaining on Earth, since the Sun will then expand beyond our orbit. Looking at the bigger picture, the universe itself is condemned to ultimate futility, on a much longer timescale of very many billions of years. It will either collapse back into a 'big crunch' or, much more likely, continue to expand indefinitely, progressively becoming cooler and more dilute until all carbon-based life must disappear from everywhere within it. In fact, every story that science can tell eventually ends in death and decay. From the point of view of physics, this is due to the second law of thermodynamics, which decrees that ultimately disorder always wins over order since there are overwhelmingly more ways of being disorderly than there are of being orderly. From the point of view of biology, an evolving world is one in which the death of one generation is the necessary cost of the possibility of the life of the next.

EVOLUTION[12]

The universe which came into being 13.7 billion years ago as an almost uniform expanding ball of energy is now a richly structured world with a great variety of life on at least one of its planets, including self-conscious human beings. Cosmic history is the story of unfolding fertility, and the processes that have brought this about are mostly of a kind that can be characterised as evolutionary. The generic character of evolutionary process lies in the interaction between two opposing tendencies which, in a slogan kind of way, can be labelled Chance and Necessity. These are slippery words which need careful definition. Chance does not stand for a series of capricious acts by the Goddess Fortuna, but it simply signifies the contingency of particular occurrences, that this happens rather than that. By Necessity is signified the lawful regularities that shape and constrain what is happening. In the familiar story of biological evolution, Chance is represented by the way in which apparently random genetic mutations give rise to new forms of life. Necessity is represented by the way in which these novel variations are sifted and preserved by natural selection, a process which has to operate in a lawfully regular environment in which there are different degrees of flourishing and rates of reproduction for creatures with dif-

12. For a discussion by a theologically aware biologist, see A. R. Peacocke, *God and the New Biology*, Dent, 1986.

ferent characteristics. In addition, for biological evolution to be possible, the laws of nature characterising Necessity have to take the very specific, 'finely tuned', form that is required if there is to be any carbon and any form of carbon-based life, as was discussed in the previous section.

Terrestrial biological history is by no means the sole example of the fertility of evolutionary process. We saw earlier how the random distribution of small fluctuations in the matter/energy of the early universe (Chance) provided the seeds from which the lawful action of gravitational attraction (Necessity) would, over time, induce the condensation of galaxies and stars. There is no analogue to natural selection in this process, but that is not essential to the general concept of evolution. Natural selection is simply a particular mechanism which happens to operate in the biological case.

The interaction between Chance and Necessity is a kind of shuffling exploration of potentiality, bringing to birth aspects of the inherent fruitfulness of the universe. The range of possible happenings that might have occurred is so vast that only a small fraction of cosmic potentiality will ever actually be realised. A particular genetic mutation turns the stream of life in a particular direction: had another mutation occurred, life would have developed slightly differently. Human beings might have had six fingers instead of five. This particularity does not imply, however, that the evolution of life is just a kind of drunkard's walk, meandering capriciously through an almost infinite space of possibility, with the implication that, if the tape of life were to be run again (as Stephen Jay Gould used to like to put it), nothing remotely like our biosphere could be expected to emerge the second time. Simon Conway

Morris has emphasised the phenomenon of convergence in biological evolution,[13] the way in which closely similar kinds of structure have emerged independently time and again. The many distinct occasions on which eyes have evolved provide a well-known example. Conway Morris believes that the range of basic structures that are both biologically accessible and functionally effective is much smaller than many palaeontologists have recognised. Possibility space is far from infinite. Of course, if the tape of life were run again, homo sapiens, in all its five-fingered specificity, could not be expected to emerge a second time, but Conway Morris thinks that it is reasonable to believe that self-conscious beings, recognisably similar to ourselves, would eventually appear.

Evolutionary process corresponds to the general scientific insight that regimes in which truly novel possibilities can emerge are always 'at the edge of chaos', that is to say, they are situations in which contingency and regularity, Chance and Necessity, interlace in delicate balance with each other. Too far on the orderly side of this frontier and things would be too rigid for anything other than more of the same to be possible. Too far on the disorderly side of the frontier, and things would be too haphazard for any novelties that did emerge to be able to persist. For example, if there were no genetic mutations, there would be no new forms of life; if genetic mutations were too frequent, no species could become established on which natural selection could bring its sifting effects to bear. The astonishingly fruitful 3.5 billion year history of life on Earth, in the course of which a world that for more than 2 billion years contained only single-celled organisms became

13. S. Conway Morris, *Life's Solution*, Cambridge University Press, 2003.

a world with persons in it, has been possible because there is just the right balance between contingent novelty and regular process acting within it.

Three further points need to be made about evolutionary processes. One is that it is as essential to understand the nature of the necessity involved (in the biological case, the environmental context and anthropic fine-tuning) as it is to understand the chance aspect (in the biological case, the nature of genetic mutations). Those wishing to emphasise what they believe to be the 'meaningless' character of evolution have too frequently been unwilling to recognise the significance of the inbuilt natural potentiality being explored. The rich diversity of terrestrial life is enabled by biochemical processes that themselves derive from atomic properties that are consequences of the physical laws of electromagnetism and quantum theory. The fundamental equations describing these laws could literally be written down on the back of an envelope, yet it is a remarkable fact that they have elephants and human beings among their astonishing consequences.

In all evolutionary explanation, a fully adequate understanding of the contextual factors is as important as understanding the genetic factors. Conventional evolutionary thinking in biology supposes that the process takes place within a context simply defined by physics and biology. However, I have suggested that the evolution of human mathematical abilities—far exceeding the simple ideas of arithmetic and geometry which have obvious everyday survival value—only becomes intelligible if that context is recognised also to include a noetic world of mathematical truths, into whose exploration our ancestors were drawn, not by survival necessity but by the lure of intellectual satisfaction.[14] Some ethical intuitions, such as kin altruism (protecting the family gene pool) and reciprocal altruism (help given in the expectation of a return), are no doubt at least partly to be understood in conventional evolutionary terms, but what about the radical altruism that leads a person to risk their life in order to rescue an unknown and unrelated child from drowning? Rather than recourse to the speculative 'Just-so' stories of evolutionary psychology would it not be sensible to consider the possibility that there is a further contextual dimension corresponding to a realm of ethical truth and value, which a theist would see as originating in intimations of the perfect goodness of the divine will?

The second point is that when processes take place at the edge of chaos, while great fertility can result, there will also inevitably be the possibility of ragged edges and blind alleys. Genetic mutation of germ cells has driven the evolution of life, but when somatic (body) cells mutate they can become malignant. The existence of cancer is part of the shadow side of biological evolution.

The third point is to deny strong genetic determinism in the development of individuals. Of course, our genetic inheritance constrains what we are and what we can do but, as is clearly shown by the differences between the life developments of identical twins, there is still much room for the formative effects of individual experience. There are only from twenty to twenty-five thousand genes in the human genome, far too few to provide a 'blueprint' for the immensely intricate structure of the human brain, with its 10^{11} neurons and

14. ER 3.

about 10^{14} synaptic connections, even when one takes into account the complex patterns of interaction between the genes. A great deal of neural structure must be formed epigenetically, that is, as a result of individual experience. A striking illustration of this fact was given when it was discovered that the part of the brain concerned with spatial relationships is unusually highly developed in London taxi-drivers, who had spent two years acquiring detailed knowledge of the vast road network of that city, necessary if they were to be licensed to ply their trade.

TIME[15]

Few experiences seem more basic to humanity than our perception of the passage of time, and few experiences are more puzzling to think about, both scientifically and philosophically. From the time of Parmenides onwards, there have been those who have maintained that the true nature of reality is timeless and that our impression of the flow of time is simply a trick of human psychological perspective. Today, this view is taken by the proponents of what is called the 'block universe', the idea that the whole of space and time treated together as a single entity (a kind of frozen block of history) is the true reality. This was a stance enthusiastically endorsed by Albert Einstein. However, many of us refuse to deny the common-sense experience of the passage of time and believe that the universe is an open realm of true becoming, so that the future is not 'already' in existence, so to speak waiting for us to arrive.

Two main lines of argument have been advanced in support of the block view. One appeals to the insights of spe-

15. FSU 7; ER 6; TCS 3.

cial relativity. Unquestionably, that theory implies the fact, well confirmed in experimental studies of the decays of fast-moving unstable particles, that different entities experience the rate of the passage of time in different ways according to their different states of motion. As the physicists say, moving clocks run slow. A second and related point is that it turns out that different observers will make different judgements about the simultaneity of distant events. The first of these two points simply refers to experienced rates of internal processes and of itself it neither supports nor contradicts the idea of the block universe. The second point, however, is often cited in support of the block view. If one observer judges that event A preceded event B, a second observer that A and B were simultaneous, and a third observer that B preceded A, does this not show that the distinctions between past, present and future are illusory? Actually, it does not. No observer has knowledge of a distant event until it is unambiguously past (technically, within the observer's past light cone). The differing judgements of simultaneity simply correspond to the differing ways in which the observers organise their descriptions of the past, and this fact can do nothing to establish the pre-existing reality of the future.

A second argument for the block view points out that physical theory contains no explicit representation of 'the present moment' in its formulation. This is certainly the case, but it can simply be held to indicate a limitation on the point of view of physics, rather than the necessary denial of a basic human experience.

The truth of the matter is that the nature of temporality, just like the nature of causality, is a metaphysical question whose answer is not determined by scientific considerations alone. It is also important to recognise that metaphysical decisions about the nature of time and the nature of causality are independent of each other. Temporality is concerned with the existential status of events (the significance to be attributed to before-and-after ordering); causality is concerned with relationships of influence acting between events. Taking the block universe view does not commit one to a belief in determinism, nor does the unfolding view of time necessarily imply a degree of causal openness to the future. The stances taken by two important thinkers of the past illustrate the point. Thomas Aquinas believed that from the standpoint of eternity God enjoyed what was effectively a block universe view of creation, perceiving it *totum simul*, all at once. Nevertheless Aquinas believed in the reality of human free choice. Such choices are not *fore*known by God, but they are all *simultaneously* present to the divine atemporal gaze. Pierre Simon Laplace, the greatest of Newton's successors, believed in a strictly deterministic universe, in which total knowledge of the present would enable complete retrodiction of the fixed past and prediction of the fixed future. Nevertheless, he thought of that deterministic world in terms of a temporally unfolding process.

Those who take the open view of the nature of time embrace the concept of the moving present, with the implication of the existence of a preferred time axis defining its nature, and they have to explain how they relate this to special relativity's ban on the possibility of discerning any preferred time axis demonstrable in terms of local physics. This ban must be observed, but nevertheless the answer to the question of the status of the moving present may be held to lie in cosmology. For the universe itself there does seem to be a special axis of time, defined in terms of a frame of reference at rest with respect to the cosmic background radiation. This is the cosmic timeframe that the cosmologists use when they give the age of the universe as 13.7 billion years. It might seem strange to assume that local experience of time could be linked with the global state of the universe, but there is another apparent example of this happening. A scientific insight called Mach's Principle points out that properties of local dynamics (technically, local inertial frames of reference) appear to be related to the fixed stars, that is to say the overall distribution of matter in the cosmos.

CONSCIOUSNESS[16]

Perhaps the most astonishing event in cosmic history that is known to us is the dawning of self-conscious life here on Earth. In our ancestors the universe had become aware of itself, and as a result the development of science became an eventual possibility. Yet consciousness is even more puzzling to understand than the passage of time—and even more undeniably a part of human experience. So far, the contribution that science can make to a fundamental understanding of consciousness has proved modest and this section will be correspondingly short, despite the importance of its subject.

A sharp tap on the head with a hammer will establish that consciousness is related to the physical state of the brain, but this fact certainly does not lead to an inevitable identification of thought with nothing more than neural processing, although it does encourage the view that we are intrinsically embodied beings. A dualist picture of human beings as incarnated souls (a spiritual entity housed in the husk of a fleshly

16. SCB 1; GH 9; ER 3; TCS 4.

body) has come for many today to seem to be an implausible stance to take. A notorious and persisting difficulty with dualism has been to understand how the two allegedly distinct substances of mind and matter could actually interact. Instead, human beings can be conceived as being animated bodies, that is to say beings for whom the mental and the material are to be thought of as complementary aspects of a single reality. Unfortunately, this kind of dual-aspect monism is as yet far from being fully articulated. In a striking phrase, Thomas Nagel called its adoption 'pre-Socratic flailing around',[17] comparing it with the views of people like Anaxagoras and Anaximander, who had the clever idea that there might be one kind of fundamental stuff out of which the variety of the world is made, but who were two and a half millennia too early to know about quarks and gluons. The pre-Socratics had to wave their hands, but they were doing so in what would eventually prove to be a very fruitful direction. The modern recognition of a duality between energy and information, which we noted earlier, has at least the air of giving some modest contemporary encouragement to taking the dual-aspect view seriously, although, of course, it can afford no more than the merest glimmer of how one might think about mental/material complementarity. It is clear that an immense expansion and enrichment of the concept of 'information' would be necessary before it became relevant to the complexity of human personhood. At present, the problem of how to speak adequately of human nature is beyond anyone's power to solve. Honesty compels us to hold fast to our basic convictions of the reality of both the material and the mental dimensions of experience, and to resist

17. T. Nagel, *The View from Nowhere*, Oxford University Press, 1986, p. 30.

succumbing to the temptation to go for a quick, oversimpli-
fied, reductionist solution, based on a Procrustean rejection
of half the evidence of experience. The physicists had to hold
on to the experimental evidence for the presence of both wave
and particle properties in light for 25 years before quantum
field theory was discovered to dissolve the seeming paradox.
Human personhood is surely immensely more subtle than
light.

Modern neuroscience is making very important discoveries about the neural pathways by which our brains process the information that comes to them from the environment, but there is an immense gap yawning between this kind
of talk and the simplest mental experiences, such as feeling
thirsty or seeing red, a gap which no one today knows how
to bridge successfully. The problem of qualia (feels) is a hard
problem indeed.

It is very interesting that scans show that certain parts of
the brain light up when certain types of mental activity are
taking place, such as scientific thinking or religious medita-
tion, but this fact simply reiterates the point that humans are
embodied beings, all of whose activities will have corporeal
counterparts, without necessarily implying the reductionist
premise that neural accounts furnish the total story of what is
going on. Observations of this kind do not themselves establish the true nature either of scientific insight or of religious
experience. Bombastic claims that consciousness is the 'last
frontier' that the heroic armies of a reductionist science are
just about to cross seem altogether empty talk.

It is just possible that the origin and nature of consciousness will never be susceptible to purely scientific treatment.
Everything else that science studies, whether matter or life,

can be treated as external to the scientist, but consciousness is intrinsically internal. No one has direct access to another's consciousness but only to their own. We do not know if we all experience red in the same way, but only that we can agree about attaching that label to the same objects, whatever our private experience of its significance may be. To note this is not to offer an argument urging giving up the scientific study of consciousness, but simply to note its possible limitations. This difficulty cannot be finessed by foolishly labelling these private experiences 'folk psychology', in an ill-judged attempt to dismiss the significance of a basic aspect of the human encounter with reality.

CHAPTER FOUR

Theology and Science in Interactive Context

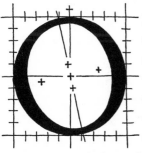NE reason why theology and science need to be in dialogue with each other is that theology asserts the universe to be a divine creation and it therefore should be gratefully attentive to all that science can tell it about that universe's nature and history. Theology might hope that from the results of this dialogue it would prove possible to discern hints of a divine Mind behind the order of the world, and a divine Purpose behind its fruitful history, even if the veiled nature of the God's presence can be expected to mean that these rumours of deity will not have the unambiguous clarity that would compel assent from those not inclined to view things in a theistic perspective. Such hints of God will not arise from mistaken attempts to provide theological answers to scientific questions, for we have very good reason to believe that scientifically stateable questions will receive scientifically stateable answers, without the need for assistance from theology. I shall later criticise the strategy of

trying to find gaps in science's understanding of natural processes into which a 'God of the gaps' could be inserted as the alleged explanation. The true Creator and Ordainer of nature must be supposed to act as much through natural processes as in any other way, and God is not to be discerned solely in the ill-understood aspects of the world. It is not the case that if natural processes did it, God had no hand in the matter.

Yet, we have good reason to believe that there are many meaningful questions that are necessary to ask but which are not scientific in their character and so cannot be addressed by an honest science. Some of these questions actually arise from the experience of doing science but are not themselves of a scientifically answerable kind. They are what may be called metaquestions, taking us beyond the self-limited confines of their scientific origin. Science has achieved its great success precisely by bracketing out questions of meaning, value and purpose and limiting itself solely to the question of the processes by which things happen. Full understanding requires a broader and deeper context than that which science on its own can provide, and theology can play a significant part in forming that wider view of reality—hence the stance endorsed in chapter 1 of seeing theology and science as being in a complementary relationship with each other. Natural theology provides a convenient setting in which to begin to explore this complementarity.

NATURAL THEOLOGY[1]

Natural theology is the attempt to learn something of God by the exercise of reason and the general inspection of the world.

1. SC 1–2; RR 6; BG 1; FSU 4.

The flow of its argument is from the consideration of the way things are to the claim that this provides motivation for belief in the existence of a divine Creator. Older styles of natural theology tended to employ the stronger language of 'proofs of the existence of God', but the modern discourse is more realistically modest in its tone. Much modern philosophical discussion of the status of human knowledge, whether in science, theology or elsewhere, has recognised that its character is best expressed in terms of the attainment of persuasive explanatory insight to which it is rational to commit belief, rather than in terms of a logically coercive understanding from which it would be wholly irrational to withhold assent. The language of 'best explanation' has replaced that of claims for the attainment of certain proof. We have already noted that, even in mathematics, Kurt Goedel showed that a degree of commitment is necessary beyond the strictly demonstrable, since axiomatised systems cannot establish their own consistency.

Natural theology claims that theistic belief affords the best explanation of certain features of the scientific exploration of the universe which seem remarkably significant in their character, but which from science's own explanatory perspective would have to be treated simply as brute facts or happy accidents. These features can be indicated by asking two metaquestions.

The first such question asks why is science possible at all in the deep way that has proved to be the case? Of course, evolutionary survival necessity can be expected to have moulded our brains so as to make them able to make sense of the world of everyday experience. However, our human ability to understand the subatomic quantum world, totally different in its character from the macroscopic world of everyday happenings and requiring counterintuitive ways of thinking for its understanding, is another matter altogether. That subatomic world has no directly discernible impact on human experience and to regard its intelligibility to us as simply the result of a happy spin-off from mundane survival necessity is a highly implausible suggestion. The fact is that the universe has proved to be rationally open to human enquiry to a very remarkable degree. And the mystery is deeper even than that, for it has turned out that it is mathematics—that most abstract of disciplines—which time and again has provided the key to unlocking the secrets of the physical universe. It is an actual technique of discovery in fundamental physics to seek theories that are expressed in terms of equations possessing the unmistakable character of mathematical beauty, a property which the mathematically minded can readily recognise and agree about. It involves such qualities as economy, elegance and what the mathematicians call 'being deep', by which they mean that extensive consequences are found to flow from seemingly simple initial definitions, as when the endless baroque complexities of the Mandelbrot set are seen to derive from a specification that can be written down in a few lines. This heuristic strategy is no mere act of aesthetic indulgence on the part of the physicists, for 300 years of enquiry have shown that it is just such mathematically beautiful theories that prove to have the long-term fertility of explanation that convinces us that they are indeed describing aspects of the way things are. In other words, some of the most beautiful patterns that the mathematicians can think about in their studies are found actually to be present in the structure of the physical world around us. A distinguished nuclear physicist, Eugene Wigner, once asked, 'Why is mathematics so unreasonably effective?' Those seeking an understanding as complete as possible must ask what it could be that links together the reason within (mathematical thinking) and the reason without (the structure of the physical world) in this remarkable way? The universe has not only proved to be astonishingly rationally transparent, making deep science possible, but also rationally beautiful, affording scientists the reward of wonder for all the labours of their research. Why are we so lucky?

It would surely be intolerably intellectually lazy not to seek to pursue this question. Yet science itself will not provide its answer, for it is simply content to exploit the opportunities that these wonderful gifts afford us, without being in a position to explain their origin. Theology, however, can step into the breach. Science has disclosed to us a world which, in its rational transparency and beauty, is shot through with signs of mind, and religious belief suggests that it is indeed the Mind of the Creator that lies behind the wonderful order of the universe. I believe that the reason within and the reason without fit together because they have a common origin in the God who is the ground both of human mental experience and of the existence of the physical world of which we are a part. The fact of an intelligible universe itself becomes intelligible when the world is seen as being a divine creation and human beings, to use an ancient and powerful phrase, as creatures made in the image of their Creator. The claim being made is not that this insight is logically incontestable and could not be denied, but that it offers the deeply intellectually satisfying best explanation of the remarkable access that science has been able to attain to the deep structure of the universe.

The second metaquestion is more specific in its character. It refers to the remarkable 'fine-tuning' of the laws of nature that we noted in chapter 3 as having been indispensable if the universe were to be able eventually to evolve the complexity of carbon-based life. This second question simply asks why is the universe so special? Scientists prefer the general to the particular and our initial expectation was that the cosmos would simply turn out be a fairly typical specimen of what a universe might be like. The insights of the anthropic principle make it clear that this is not the case. The laws of nature that operate in our universe take the very precisely defined form that is required for the possibility of carbon-based life. Clearly those seeking as full an understanding as possible will not be content to treat this as another monstrously happy accident. The religious believer will see cosmic fine-tuning as an endowment of potentiality given by the Creator to creation in order to bring about a fruitful history that fulfils the divine Purpose. Disliking this threat of theism, some scientists have proposed the alternative concept of the multiverse,[2] the notion that our world is just one member of a vast collection of different worlds, all with different laws and circumstances and all, except our own, inaccessible to our observation. If this were the case, then our particular universe might simply be the one which, just by chance, happened to have the right laws and circumstances to allow carbon-based life to develop—a kind of winning ticket in a gigantic multiversial

2. See R. D. Holder, *God, the Multiverse and Everything*, Ashgate, 2004, ch. 7.

lottery. This suggestion is not only excessively ontologically prodigal (William of Ockham must be turning in his grave!) but also not even clear without further argument that it attains its purpose. The existence of even an infinite number of universes would not of itself guarantee that one of them had life-generating capacity, as if all possibilities would have to be present in that vast ensemble. After all there are an infinite number of even numbers but none of them possesses the property of oddness. There are some highly speculative ideas in physics, such as string theories, which might give some encouragement to the hypothesis of a multiverse, but their conjectural nature and lack of observational support must mean that sober assessment of the notion of the multiverse has to be that it is a metaphysical proposal, just as the existence of a divine Creator is a metaphysical proposal. The latter also does further explanatory work, however, such as that relating to the deep intelligibility of the universe (which we have just discussed) and giving the reason for the widespread human testimony to encounter with a sacred dimension of reality. The multiverse, however, only seems to do one thing, to explain, or to explain away, anthropic fine-tuning. There is a cumulative case for theism which does not have a counterpart in belief in the multiverse. If the multiverse is to be truly explanatory, it needs some extra collateral support since without it, attempting to hide a significant particularity in a hypothetical infinity of other possibilities would not amount to an explanation at all. Anything could be 'explained' away in that cavalier fashion. Any needle of significance could be hidden in the conjecture of a sufficiently capacious haystack.

Even if the arguments presented so far are accorded

their most persuasive force, one must confess that they lead only to a rather thin idea of God, conceived as something like the Cosmic Architect or the Great Mathematician. This is scarcely surprising, since natural theology only looks to a limited kind of general experience to motivate its understanding, and consequently it can only offer limited insight into the nature of deity. By itself it is as consistent with the spectatorial God of deism, who simply decreed the order of creation and then watches from a detached distance as its history unfolds, as it is with the providentially interactive God of theism. Natural theology leaves unanswered such questions as whether God cares for individual creatures. Answering these questions will require the consideration of more personal and particular claims of encounter with divine reality, of the kind that Christian theology, for example, regards as being revelatory disclosures of the divine nature. The issue of revelation is one to which we shall come later.

An alternative theological stance in relation to the issues of cosmic intelligibility and anthropic fine-tuning could be framed in terms of a *theology of nature*. The difference between this strategy and natural theology lies in the direction of the flow of the argument. The latter seeks to move from scientific insight to motivation for belief in the existence of a Creator, but a theology of nature is less ambitious, simply positing the existence of God from the start of the argument and seeing it as the source from which flows a satisfying understanding of the scientifically discerned character of the world. The more modest character of this approach means that it not only sees the Mind of God as the origin of the order of the universe and the divine Purpose as being fulfilled in its fruitful history, but it can also offer an enhanced range of insights expressing further consonance between theology and science, without claiming that they constitute grounds for an apologetic argument for belief in the existence of the Creator. For example, from the standpoint of the Christian trinitarian understanding that the nature of the Godhead is constituted by the mutual relationship of love between three divine Persons,[3] the role of relationality, discovered by science to exist within created reality and noted in the previous chapter, becomes readily understandable. We would expect that the nature of the Creator finds some pale reflection within creation. Yet no one could pretend to argue in the reverse direction, trying to move from quantum entanglement to the triune nature of God.

A theology of nature must also be expected to have much to say about the scientific discernment of the widespread role of evolutionary processes in the history of the cosmos, and this is an issue that will be considered in the section which follows.

Finally in this section, we should note that, lying outside the dialogue between science and theology but having connections with both natural theology and a theology of nature, are questions about the existence and source of moral and aesthetic values. A theist will see human moral knowledge as originating in intuitions of the good and perfect will of God, and aesthetic delight as a sharing in the Creator's joy in creation. Theism thus provides a satisfyingly integrated understanding of the origin of a wide range of basic human experience. The concept of God as the ground of value offers insights of great significance, but not ones to be pursued spe-

3. See, for example, C. M. Lacugna, *God for Us*, HarperSanFrancisco, 1973.

cifically in the context of the interaction of theology and science.

CREATION, EVOLUTION AND EVIL

The doctrine of creation, properly understood, is concerned with the question of why things exist and not simply with how things began. The question it addresses is 'Why is there something rather than nothing?' not merely 'Who lit the blue touch paper of the big bang?' In considering creation, theology has to take account of all that science can tell it about what exists and what the character of its history has been, for God is as much the Creator today as 13.7 billion years ago. While the physics of the early universe is certainly interesting, it holds no unique significance for theology. If the speculation of a steady-state universe, existing over limitless periods of time, had proved to be the correct cosmological picture, theology could have lived with that theory, even if some details of its discourse on creation would have required modification. The fundamental belief that creation is held in being by the will of its Creator would not have been threatened.

The preceding discussion of natural theology has already presented motivations for the metaphysical beliefs that a divine Mind lies behind the order of the universe and a divine Purpose behind its fruitful history. We have also seen in chapter 3 that the processes, which over 13.7 billion years have turned the initial ball of energy into a world with persons in it, have been evolutionary in character, whether one is thinking of the formation of the stars and galaxies or the increasing complexification of life on Earth. We also noted in chapter 2 that from the first there were religious people who gave a welcome to Darwin's evolutionary ideas, in a way that was aptly summarised in Charles Kingsley's pregnant remark that an evolving world should be seen theologically as a creation in which creatures are allowed 'to make themselves'. It is time to give this idea more detailed attention.

The first point to make is that the idea of evolution clearly encourages thinking of creation as involving an unfolding process and not simply as a single initiating act. Evolution's emphasis on the role of developments taking place over vast tracts of deep time lays stress on the creative significance of temporality. Time is not simply the index of when things happened, but it has a fundamental categorical status, reflecting the fact that the character of the present is constituted by the events of the past. This understanding has led to theological recognition that there are two distinct aspects of the concept of creation.[4] One is the concept of creation out of nothing, *creatio ex nihilo*. This does not mean the odd idea that God made the world out of some sort of curious 'stuff' called *nihil*, but that the world, unlike its Creator, does not possess being in itself, but it is held in existence, moment by moment, by the divine Will alone. The second and complementary aspect, strongly encouraged by evolutionary insight, is *creatio continua*, the unfolding creative process by which potentiality is continuously being transformed into actuality. It was just this second thought that led Aubrey Moore to say that Darwin in the guise of a foe had done the work of a friend. Moore had recognised that evolutionary ideas encouraged the replacement of the notion of a distant Creator, with

4. SC 4; SCB 4; S as T 4. A. R. Peacocke, *Creation and the World of Science*, Oxford University Press, 1979.

its dangerous threat of degenerating into mere deism, with the picture of a God continually at work in and through the processes of created nature. The Creator is not to be thought of as acting through acts of occasional interference in cosmic history, but is at all times the God of the whole show. To suppose the contrary and to look for signs of a 'God of the gaps', only discernibly active in processes currently not well understood scientifically, would be a bad theological mistake as well as a disastrous apologetic strategy. Such a Creator could be no more than a kind of 'Cheshire cat deity', continually disappearing from sight with the advance of scientific knowledge, continually driven to take refuge over the next intellectual horizon.

Christian theology has to steer a course between two unacceptably extreme accounts of God's relationship with creation. One is that of the indifferent deistic Spectator, who having set the world going just stands back and watches it all happen. Equally unacceptable would be the picture of the Cosmic Tyrant, whose creation is a puppet theatre in which the Grand Puppet Master pulls every string. Neither of these accounts is consistent with the Christian belief that God is the God of love, for the gift of love must always include some due degree of freedom granted to the objects of love, as well as a continuing interactive concern for creatures. One might dare to say that an evolving creation, in which creatures are allowed to be themselves and to make themselves, is a more fitting creation for such a God than a ready-made world would have been. Yet it is also a world with an inescapable shadow side to its history. The shuffling explorations of potentiality that result from the interplay between Chance and Necessity at the edge of chaos will not only bring great fruitfulness to birth but will also inevitably have ragged edges and sometimes lead to blind alleys.

We saw that genetic mutation has been not only the engine driving the remarkable 3.5 billion year history of the development of terrestrial life but also a source of malignancy. One cannot have the one without the other. This insight offers theology some help as it struggles with its most perplexing problem, the existence of natural evil and suffering in a world believed to be the creation of a good and powerful God.[5] It is important to understand what is meant by calling God 'almighty'. It does not mean that God can do absolutely anything without any restriction whatsoever. Of course, nothing restricts God from the outside, but there are the internal constraints arising from divine consistency. God can do what God wills, but that will only be what is in accord with the divine nature itself. The rational God cannot decree that 2+2=5. I have already suggested that the God of love cannot act as a domineering Cosmic Tyrant, exercising tight control over every detail of creation. The gift of freedom to human beings implies that God cannot at the same time ensure that all their choices are for the good. In relation to moral evil—the chosen cruelties and neglects of human beings—this realisation is the basis of the so-called free-will defence, claiming that a world of freely choosing beings is a better world than one populated by perfectly programmed automata, despite the disastrous choices that these free beings may sometimes make. After the twentieth century, with its wars and genocides, this is not an easy assertion to make, but I believe nevertheless that it ex-

5. SP 5; ER 8.

presses a truth. I have suggested that in relation to physical evil—disease and disaster—there is a complementary free-process defence,[6] claiming that a creation in which creatures are allowed to behave in accordance with their natures and to make themselves is a better world than a ready-made creation would have been, despite the former's shadow side.

At the heart of the free-process defence lies the conviction that the disease and disaster present in nature is not gratuitous, something that a Creator who was a bit more competent or a bit less callous could easily have remedied, but it is the inescapable consequence of the nature of a freely fertile world.[7] We tend to think that if we had been in charge of creation, frankly, we would have done it better. We would have kept the good things (beauty and fruitfulness) and eliminated the bad things (disease and disaster). However, science makes it clear that this is not possible, for the way the world works is a kind of package deal. Its processes are so intrinsically entangled (remember the operation of Chance and Necessity entwined at the edge of chaos) that it is not possible to separate the good from the bad, retaining the one and discarding the other. The ambiguous character of genetic mutation illustrates the point. Another example is provided by the destructive nature of tsunamis. They occur because of undersea earthquakes, which in turn are caused by the slipping of tectonic plates. One might have supposed that it would have been better for the Creator to have arranged for the Earth to have a solid crust, but this is not the case. The gaps between the plates allow mineral resources to well up from within the Earth and replenish its sur-

6. SP, pp. 66-7.
7. See also, C. Southgate, *The Groaning of Creation*, Westminster John Knox, 2008.

face fertility, a process which is vital to maintaining the long-term flourishing of life upon it.

Two theological concepts are important in the discussion of the doctrine of creation. The first is the recognition we have been exploring that the act of creation is an act of self-limitation on the part of the Creator in bringing into being creatures who are truly other and who are truly allowed to be and to make themselves.[8] This creaturely freedom is permitted by a voluntary curtailment by the God of love of total control over all that happens. The theological word for this self-limitation is kenosis (the Greek for emptying), a concept first stated in relation to the Christian understanding of the incarnation of the divine Word in human form (Philippians 2:7) and eventually recognised as of wide significance for the whole of God's relationship with creation. Much contemporary theological thinking lays emphasis on the kenotic character of divine creativity, and we shall return to this idea in later discussion.

The second concept is the idea of theistic evolution, accepting all that science can tell us about cosmic and terrestrial history, but setting the story in the context of the unfolding of God's purposes. Not only does this involve recognising that the Creator of the world is acting in a general way through the natural processes which God has ordained but it can also include the possibility of specific divine providential action, exercised not by occasional interference, but continuously within the open grain of nature. How this latter possibility might relate to scientific insight is the topic to which we must now turn.

8. See essays in J. C. Polkinghorne (ed.), *The Work of Love*, SPCK/Eerdmans, 2001.

DIVINE PROVIDENCE[9]

Christian theology requires an understanding of the Creator's relationship with creation that can accommodate belief in divine action within history, without denying the reality of the gift of a due degree of freedom for creatures to behave in accordance with their natures. We have seen that this divine gift of creaturely freedom is the basis for theological response to the problem of evil and suffering, expressed through the free will defence and the free process defence. God is to be thought of neither as a deistic Spectator of the history of creation nor as a Cosmic Tyrant keeping all history within tight divine control. How the balance between creaturely action and divine action is actually struck is the classic theological problem of grace and free will, now written cosmically large. Had science established the causal closure of the world on its own terms, based solely on the exchange of energy between constituents, there would have been no such balance to strike. All would have been physically determined from the bottom up, and the Creator, after having brought the world into being, would have had no further role to play other than keeping it in being while watching what happens in consequence. No form of specific divine influence, responsive to events, could in that case be operating in the course of the universe's unfolding history. The rise of deism in the eighteenth century was undoubtedly influenced by belief that the post-Newtonian science of the day implied a clockwork universe. We saw in chapter 3, however, that the existence of intrinsic unpredictabilities, together with the inescapably metaphysical character of any ultimate account of causality, imply

9. SP 1–3; SCB 4; BG 3; FSU 6; S as T 3.

that this conclusion is by no means required by what an honest science can actually contribute to an understanding of the nature of causality.

Realisation that this is the case has placed the topic of divine providential action high on the contemporary agenda of those working at the interface between science and theology.[10] A way of approaching the matter can be developed by recognising three stages of increasing complexity present in an account of created nature. The first stage simply looks at the self-organising powers found to be possessed by physical dissipative systems, treated in the manner set out in chapter 3. There this led to the hypothesis of the existence of holistic laws of nature, complementing the constituent laws that have been science's traditional account of causal structure. The picture proposed was based on the hypothesis of the operation of a top-down holistic causal principle, for which the name of 'active information' was suggested, with 'information' being taken in this context to mean the specification of dynamical patterns of large-scale orderly behaviour. It was further suggested that this account of causality might give a *glimmer* of understanding of how there could be a dual-aspect account of psychosomatic human nature, with the mental and the material united in an intimate complementarity, a strategy that would offer to accommodate the basic human experience of the exercise of willed action. This last step of accommodating human agency constitutes the second stage of the discussion. Of course, for this move to be successful, 'information' would

10. See the series edited by R. J. Russell et al., Vatican Observatory/CTNS: *Quantum Cosmology and the Laws of Nature*, 1993; *Chaos and Complexity*, 1995; *Evolutionary and Molecular Biology*, 1998; *Neuroscience and the Person*, 1999; *Quantum Mechanics*, 2001.

need to become a concept generalised and enriched to an immense degree beyond the comparative banalities of dynamical patterning. Embracing this conjectural insight about human agency, would then lead naturally to a third stage of the argument, seeking to understand divine providential action as also being exercised within the open grain of created process, by the input of pure information (that is, without an accompanying need for physical embodiment).[11] Clearly great speculative boldness is being exercised in this project, but its account bears some modest analogical kinship with the theological idea of the divine Spirit at work 'on the inside of creation'.

These ideas are consistent with the possibility of the input of information without a necessary energetic accompaniment, since they are compatible with the ontological interpretation of both kinds of intrinsic natural unpredictability discussed in chapter 3. The different paths traced by a chaotic system through its strange attractor have the same energy content and differ only in the pattern of its flow. The same is true, within the limits of uncertainty, of the different outcomes of quantum measurement events.

Yet one must recognise that it is beyond the power of contemporary thought to be able to offer an adequately articulated account of the exercise of agency, whether human or divine. All that is being claimed in the argument is the 'defeat of the defeaters', the demonstration that the insights of an honest science can be taken with the seriousness that they deserve, without necessitating rejection of belief in the exercise of human willed agency or the operation of divine provi-

11. BG 3.

dence. The gain attained by this discussion is modest but vital, both for anthropology and for theology.

In the discussion of providential agency, different people have laid stress on different aspects of intrinsic unpredictability in their attempts to speak of the 'causal joint' by which this action might be effected. Some have emphasised quantum physics,[12] supposing that God may act principally by determining all—or perhaps only some of the most potentially significant—quantum events, while keeping this carefully hidden by ensuring that the overall statistical distribution of outcomes remains compatible with the laws of quantum mechanics, for these too are expressions of the Creator's will. For such action to be manifest in the macroscopic world of historical events, these microscopic quantum outcomes would have to be promoted in some way to a much-enhanced level of consequence. Our scientific lack of understanding of how the quantum world and the everyday world are related to each other (illustrated by the difficulties of quantum chaology discussed in chapter 3) makes this a difficult issue to tackle. Moreover, quantum unpredictabilities are associated with measurements (not necessarily acts of human observation, but any macroscopic registration of a signal of the state of a quantum system). These only occur from time to time, so that divine providence exercised in this manner would be episodic and have a rather staccato character.

The approach I have taken, laying emphasis on the role of ontologically interpreted chaotic unpredictabilities, is also not without its difficulties. The processes appealed to certainly have direct macroscopic consequences. However, there

12. See N. Murphy, R. J. Russell and T. Tracey in note 10.

is currently no fully articulated detailed theory that expresses the necessary modification of Newtonian determinism. It is an entirely reasonable assumption that the true theory would have properties analogous to aspects of classical chaos, such as the existence of strange attractors, but this has not been demonstrated conclusively.

Some have sought to avoid these detailed problems by appealing to a generalised notion of top-down causality, perhaps exercised by the Creator in the ultimate fashion of a divine influence acting on the created universe taken as a whole.[13] However, causality certainly seems to be a zero-sum game and unless there is a degree of openness present in the operation of bottom-up causality, there would not be room for top-down influences of any kind to be at work.

Another suggestion has been that perhaps God interacts with human minds alone.[14] However, this is only a useful possibility to consider if one takes a dualist view of human nature. Otherwise, one must recognise that even God cannot interact with a human mind without interacting in some manner with a human brain. Moreover, this approach implies that for most of the history of creation, before the appearance of hominids, the Creator was an inactive spectator.

Yet another approach rejects altogether the search for some kind of causal joint by which providential agency might be exercised. Instead it appeals to an idea, found, for example, in the writings of Thomas Aquinas. The Creator is held to be so different from creatures that any search for an analogy between their different modes of agency is believed

13. A. R. Peacocke, *Theology for a Scientific Age*, SCM Press, 1993.
14. D. J. Bartholomew, *God of Chance*, SCM Press, 1984.

to be woefully mistaken. God acts through a primary causality that is wholly different from the secondary causality of creatures. This divine primary causality is believed to be at work in and under the secondary causalities of creatures in an ineffable manner not capable of further explanation. One may well feel that this is an unsatisfactorily fideistic account, which simply ignores significant questions that have to be addressed. It is largely motivated by a desire not to speak of providential action as if it were just another cause acting among other causes, with God an invisible Agent acting alongside creaturely agents. I would defend the causal joint approach by replying that it is another example of the kenotic nature of the act of creation that the Creator has condescended to act in this manner, choosing to operate providentially within the open grain of created reality.[15]

The fact of the matter is that a full understanding of the exercise of any form of agency is a task beyond our contemporary capacity to attain. It is very unlikely that either human agency or divine providence is exercised solely through processes either at the quantum level or at the chaotic level. Doubtless the causal structure actually involved is much more complex and entangled, involving many levels in a manner that is beyond present ability to analyse and understand. The discussion of intrinsic unpredictabilities that we have largely been pursuing would have been hopelessly over-ambitious if it had ever been thought capable of yielding a fully adequate theory of agency. I have suggested that what it does achieve is much more modest, but still valuable in showing that physics does not rule out either human belief that we are creatures able

15. Polkinghorne in note 8.

to exercise a degree of freely willed agency or religious belief that God is no deistic Spectator but is providentially active within the unfolding history of creation. An inability to offer a full scientific explanation of how things happen is no reason to deny that the fact of their happening. The story of wave/particle duality makes this clear enough within physics itself. It was 25 years before physicists gained an understanding of this apparently paradoxical behaviour, but in the meantime they had to hold on to what experience had taught them, however strange it seemed to be. The nature of agency is surely a much deeper problem than the nature of light and we should continue to struggle with it, even if the timescale for progress is likely to be long.

Two final points need to be made about the understanding of providence pursued in this section. If the regimes of intrinsic unpredictability are indeed the right direction in which to wave our hands in presocratic flailing around in search of an understanding of agency, then it will not be possible to escape an irreducible degree of cloudiness in any attempt to analyse what is going on in the world. Completely detailed clarity will not be attainable. It will not be possible to take occurrences apart and say that natural process did this, human will did that, and divine providence did a third thing. Causes will be so mutually entangled within the veil of unpredictablity that such a clear itemisation will not be feasible. The acts of providence may be discernible by faith, but they will not be demonstrable by experiment. A person sitting by marshy ground may observe the advance of a party of fugitive slaves, hotly pursued by a band of soldiers. As the slaves approach the marsh, a strong wind arises, driving back the waters and letting them get across. As the soldiers enter the marsh, the wind drops, the waters return and they are caught in the mud and drowned. No one can force that observer to see more than a remarkably fortunate coincidence, but no one can forbid the fleeing Israelites to see this as a great act of divine deliverance from a life of slavery in Egypt.

The last point to make is that the stronger the account one is able to give of divine action in history, the stronger become the problems of theodicy. We have argued that the response must lie in a recognition that the divine gift of freedom to creatures is a real one so that, though God is party to all events in the sense that if the creation were not held in being there would be no events at all, yet not all that happens is in accordance with the positive divine will in a world in which creatures are allowed to be themselves and to make themselves.

PRAYER[16]

There are many different ways of praying, ranging from silent contemplation to active petitionary prayer. I believe that many scientists pray without being aware of it, for their experience of wonder at the marvellous order of the world is surely to be understood theologically as an act of praise of the Creator. Nevertheless, it is petitionary prayer that raises most questions for the dialogue between science and theology. If God were just the god of deism, one might well contemplate with awe and gratitude the deep order that the universe had been given by its Creator, but one would not expect from such a deity any particular response to individual needs and circumstances. All that could be hoped for would be that the nature

16. SP 6.

of the world that God had set in motion would be such that the outcomes of its independent processes would tend, in general, to be constructive rather than destructive.

However, we have seen that taking science seriously does not imply a necessary denial of the reality of divine providential action. The interactive God of theism is one to whom it is possible to address petitions for particular outcomes in particular circumstances. But then another problem immediately arises. If the God of theism is indeed an active participant in the unfolding history of creation and not merely a passive spectator, why is it necessary to *ask* for things to happen? God knows far better than we do what would be best, so why does God not just get on with bringing it about? What are we doing when we pray? It does not make theological sense to suppose that we are making such a fuss that God is persuaded to do something that God would not have bothered to do without our importunity. Nor are we drawing the divine attention to something that otherwise might have been overlooked, or suggesting a cunning plan that God would not have thought of without our help.

I think that there are at least two things happening in petitionary prayer. One is the offering of human wills to God to be used in bringing about the fulfilment of the divine purpose. While God has reserved to the divine will a significant degree of providential power to bring about the future, creatures have also been given lesser powers of agency to affect the future. There is a genuine instrumentality in prayer because things become possible when human and divine wills are aligned with each other which would not be possible if they were at cross-purposes. Using a scientific metaphor, one may say that prayer seeks a laser-like coherence between divine and human wills. Laser light is powerful because it is what the physicists call coherent. All the waves are in step, so that all the crests coincide and add up, and all the troughs coincide and add down, yielding maximum effect. In incoherent light, crest and trough can coincide and cancel each other out. Prayer is seeking an act of laser-like coherence in human-divine co-operation. In the Gospels, Jesus' acts of healing require the trustful collaboration of those seeking healing. We are told that, when he visited his home town, he was met with such unbelief that 'he could do no act of power there' (Mark 6:5).

If this picture of petitionary prayer is right, then there are implications that flow from it. One is that prayer is not a substitute for action but a commitment to action. Those who pray for their neighbours must be prepared to go and help their neighbours. Another consequence is an explanation of the widespread conviction that it is a good idea to have many people praying for the same thing. This is not because there are more fists beating on the heavenly door to attract the divine attention, but because there are more wills to be aligned with the divine will in bringing about the best outcome. Yet, having said all this, one must also acknowledge the mystery of individual destiny. A person is gravely ill and the Church prays for that person's healing. The wholeness that is being sought may come in the form of physical recovery, but it might also come in the form of a peaceful and trusting acceptance of the imminent destiny of death. No one can say beforehand what form the healing will take, and afterwards only those intimately involved can say whether it was received.

The Oxford philosopher, John Lucas, has drawn our attention to a second thing that he believes is happening in petitionary prayer. He sees prayer as the opportunity to commit ourselves to what we really want, to lay before God our heart's desire. Lucas says, 'The mere fact that we want something is a reason, though not a conclusive reason, for God giving it to us'.[17] Prayer is not the frivolous filling in of blank cheques given us by a heavenly Father Christmas, but neither is it a non-negotiable transaction in which our role is passive and unquestioning acceptance of what is happening. In the Gospels, when Jesus meets a blind man outside of Jericho, the latter is called upon to answer the question 'What do you want me to do for you?' with the public response 'My teacher, let me see again' (Mark 10:51) before he is actually healed. This commitment to what we really want removes petitionary prayer from the triviality of a wish list and makes it a deeply serious activity.

Can a scientist really pray, knowing all the he or she does about how the universe works? The discussion of divine action in the preceding section enables us to answer 'Yes, a scientist can' (and in fact many do). However, there are domains of regularity as well as domains of unpredictability — clocks as well as clouds one might say — in the world that science describes. The Sun is going to rise tomorrow, summer will be succeeded by winter. Theologically these regularities have been interpreted as signs of the faithfulness of the Creator. This means that the scientist will recognise that there are some things that it is not reasonable to pray for. Long ago, the third-century theologian Origen said that one should not pray for the cool of spring in the heat of summer, tempting though this must sometimes have seemed in his native Alexandria.

17. J. Lucas, *Freedom and Grace*, SPCK, 1976, pp. 71–2.

MIRACLE[18]

Providential interaction within the open grain of created nature is one thing, but what about miracle, the claim of radically unnatural divine acts such as the resurrection of Jesus Christ? Although David Hume notoriously defined a miracle as a violation of a law of nature and considered those laws to be so surely known that no evidence of the miraculous could ever be strong enough to encourage belief in its happening, the theological understanding of miracle adopts a different approach. It regards miracles in the way that they are described in the Gospel of John, as being 'signs', that is to say, events that manifest with specific clarity some particular aspect of the divine will and nature that is normally veiled from clear sight. Miracles are not arbitrary divine actions but events of deep disclosure.

Some of these disclosure events might be manifestations of natural human powers possessed to a pre-eminent degree by a particular person and serving as a signal of that person's special significance in the divine economy. It is possible to understand some of the healing miracles of Jesus in this way. Other events might be considered as signs manifested through a providentially caused coincidence in the timing of their occurrence. Some of the Gospel nature miracles, such as the stilling of the storm on the Lake of Galilee, might be understood in this manner. But there are other claims of the occurrence of miracles that refer to events that are radically unnatural and do not lend themselves to any kind of naturalising explanation. Examples would be turning water into wine and, above all, the central Christian miracle of the resurrection of

18. SP 4.

Jesus from death to a life of unending glory. These would be a form of divine action quite different from the kind of providential action within the open grain of nature that was discussed in an earlier section. No one could suppose that the resurrection took place through some clever exploitation of the openness associated with quantum or chaotic unpredictabilities. These unnatural events, which one might call the radically miraculous, are among the signs that carry the most intense symbolic significance. This has led some to question whether they may not simply be symbolic stories without a necessary anchorage in actual historical happenings. This suggestion requires consideration case by case, and in the following chapter I shall give reasons for strongly rejecting the suggestion in the case of the resurrection.

Are claims for the radically miraculous to be seen as points of absolute conflict between science and theology? The issue is not as straightforward as it might seem. Science is concerned with what usually happens so that, for example, all it can say about the resurrection is that usually dead people stay dead, a fact as clearly known in the first century as it is today. However, if there is a God who is the Ordainer of the order of nature, then the Creator is not to be denied the possibility of exercising divine power in a manner that manifests deep aspects of that nature not discernible through normal experience. Divine action of this kind would then result in consequences lying beyond the limits of science's extrapolated expectation of normality. Yet this divine power will not be exercised in an arbitrary fashion. The real problem of the possibility of radical miracle is not scientific, but the theological issue of divine consistency. It is *theologically* inconceivable that

God is a kind of show-off celestial Conjurer, who does something today that God did not think of doing yesterday and will not bother to do again tomorrow. Divine consistency, however, is not to be thought of as like the unrelenting regularity displayed by the law of gravity. Rather, it lies in perfectly appropriate particular action in relation to specifically particular circumstances. The Creator is not condemned never to do anything radically new, but equally God does not do new things for trivial or capricious reasons. Theology uses personal language about deity—calling God 'Father' rather than 'Force'—precisely because it believes God does particular things in particular circumstances, even unprecedented things in unprecedented circumstances. If it is true that God was present in Jesus Christ in a unique way, then it is a rationally consistent possibility that the new regime that Jesus represented would be accompanied by new phenomena, even to his having been raised from death to an endless life of glory. Equally, of course, if he was so resurrected, that is surely a sign of his unique status. There is a degree of inescapable circularity here, comparable to the circular interaction in science between theory and experiment that we noted in chapter 1. Radical miracles are to be understood as radical signs, conveying hitherto unknown depths of understanding of the nature of the consistent divine will and purpose.

The point being made can be illustrated by a scientific parable. One of the most frequently verified laws of nature must surely be Ohm's Law relating current, voltage and electrical resistance. Enunciated by Professor Ohm in the early nineteenth century, it has subsequently been checked in school laboratories by generations of students. Yet in 1911, the Dutch physicist, Heike Kamerlingh Onnes, discovered that for some metals at very low temperatures Ohm's Law no longer holds. Resistance vanishes and a current will keep circulating without an electromotive force to drive it round. He had discovered the phenomenon of superconductivity. The laws of physics had not changed at the transition temperature (the point at which superconductivity sets in), but the consequences of those laws changed radically. In this story from physics, there is a continuous consistency of underlying law in the face of an apparently astonishing discontinuity of superficial outcome.

It is in a similar way that theology must seek to understand the radically miraculous. This approach has to be pursued on a case-by-case basis, for there can be no general theory of unique events. The next chapter will give an example of this rational strategy in relation to the resurrection.

GOD AND TIME[19]

Classical theology, such as that expressed in the thought of Thomas Aquinas, saw God as wholly outside of time, looking down from eternity onto the span of created history, which is perceived by the atemporal divine gaze *totum simul*, all at once. We have already noted in chapter 3 that this concept did not threaten the freedom of creatures since it involved no divine *fore*knowledge of their actions. Everything that happens is simultaneously contemporary to the timeless God inhabiting eternity. We also noted that, if this is true, it implies that God knows creation in the form of the block universe. Since God surely knows things as they truly are, this in turn

19. SP 7; FSU 7; ER 6; TCS 3.

would seem to imply theological endorsement of the block universe account of spacetime properties, although this is not a conclusion one often finds expressed in the theological literature.

In chapter 3, however, I gave scientific and metaphysical reasons for rejecting the block universe in favour of seeing the cosmos as a world of unfolding becoming, in which the future arises in novel and open ways from the past. If this is correct, then the principle of true divine knowledge surely implies that God knows creation in its actual becomingness, not only knowing that events are successive but also knowing them according to their nature, that is in their succession. This true divine engagement with time then leads to the concept of a dimension of temporal experience present in the divine nature itself. Of course, no theist can believe that God is simply temporal, in thrall to time in the way that creatures necessarily are. There must be an eternally steadfast and unchanging dimension in the divine nature also. In other words, one is led to a dipolar concept of God, with both an eternal pole and a temporal pole present within deity. Two arguments of classical theology were opposed to such an idea. One is that of divine simplicity, the claim that God could have no parts since that would imply an unacceptable dependence of deity upon those parts. I think that the answer lies in recognising the perfect mutuality and integration of the parts, much in the way that Christian Trinitarian theology recognises the essential unity of the three divine Persons. The other objection stemmed from the classical conviction that the Perfect had to be the Immutable in every respect. It was as if the point of perfection was a unique metaphysical peak from which to move in any way would be to decline. I think that we can conceive of the idea of dynamical perfection, consisting in totally appropriate relationship to a changing creation, without prejudice to the eternally steadfast character of the divine love and faithfulness that always motivates that relationship. Surely the Creator can be supposed to have related to the universe when it was a quark soup in a different way than today when it is a world with persons in it.

Process theology, derived from the thought of A. N. Whitehead,[20] asserts divine temporal dipolarity of this kind, but it supposes it to be imposed upon God by metaphysical necessity. The open theism that I am exploring here rejects the notion of such necessity and regards the divine relationship to temporality as being freely assumed in an act of kenotic condescension by the Creator on bringing into being a temporal creation. Orthodox theology speaks of the divine energies, active manifestations of God's glory beyond the inner being of the Godhead itself. It envisages these energies as eternal, even 'when' there is no creaturely 'beyond' to which they are manifested. Alternatively, one might speculate that the divine energies correspond to the engagement of the temporal pole of deity with creation once the latter is in being and endowed with temporality.

The idea of a God both eternally steadfast and unchanging in love and also interacting with and responding to the flux of history is certainly the kind of God boldly portrayed in the pages of the Bible. The Hebrew scriptures, in their anthropomorphic way, even dare to speak of God changing the divine mind in response to events (for example,

20. A. N. Whitehead, *Process and Reality*, Free Press, 1978.

2 Kings 20:1–11). Pursuing the issue, from the scientific point of view one might go on to ask what is the frame of reference which defines the divine time axis? The natural answer would be the same cosmological frame that was discussed in chapter 3 in relation to creaturely experience of the present moment. In any case, for an omnipresent Observer there are no distant events and so God knows everything that happens as and when it happens.

The dipolar view of divine eternity/temporality in relation to a created world of true becoming would seem to carry the implication that even God does not yet know the unformed future. This is no divine imperfection, since the future is not yet there to be known. God possesses what philosophers call a current omniscience (knowing now all that can be known now) but not an absolute omniscience (knowing all that eventually will be knowable). Acceptance of this self-limitation is yet another kenotic act by the Creator on choosing to bring into being a temporal creation. Of course, God can discern the way that history is moving in a manner that is far beyond the power of any creature. In that sense, God is not in thrall to time as we necessarily are. God saw that Pharaoh was not going to come from Egypt to deliver Jerusalem from Nebuchadnezzar and so God inspired Jeremiah to warn the King of Judah that he should make the best terms he could with the Babylonians. Yet we do not need to suppose that God saw beforehand every detail of the burning of the Temple.

These issues are certainly contentious and there is no unanimity in the theological community about them. Nevertheless, in the context of science a dipolar understanding of divine eternity/temporality is very persuasive.

ESCHATOLOGY[21]

We noted earlier that every story told by science ends eventually in decay and futility. All human beings must die and ultimately the universe itself will die as increasing cold and dilution lead to carbon-based life disappearing everywhere within it. If the scientific story, with its tale of the final triumph of disorder over order, were the only story to be told, that would present a serious challenge to theological belief in the world as a meaningful creation. The distinguished physicist and atheist, Steven Weinberg, notoriously said that the more he understood the universe the more it seemed to him to be pointless.[22] Theology has its own and different story to tell, however, which is based on belief in the enduring faithfulness of God. While there is no natural expectation, of a kind that science could offer us, of there being a destiny beyond death, either for creatures or for the universe itself, there is every theological expectation that the last word does not lie with death but with the ultimate reality of God. Jesus made just this point in an argument he had with the Sadducees about whether there is a destiny beyond death (Mark 12:18–27). He reminded them that at the burning bush God was declared to be the God of Abraham, Isaac and Jacob. Jesus commented that 'He is the God not of the dead, but of the living'. In other words, if the patriarchs mattered once to God, as they obviously did, then they must matter to the faithful God for ever. Abraham, Isaac and Jacob were not cast aside at their

21. SP 9; SCB 9; GH; ST 6; ER 10; TCS 7; see also J. C. Polkinghorne and M. Welker (eds), *The End of the World and the Ends of God*, Trinity Press International, 2000.

22. S.Weinberg, *The First Three Minutes*, Andre Deutsch, 1977, p. 149.

deaths, like broken pots thrown onto a rubbish heap after they had served their purpose, but they continue to live before the God whom they served. Science, which can only tell the 'horizontal' story of unfolding present process, is not in a position either to deny or confirm this 'vertical' story of divine faithfulness.

Theology's account of ultimate destiny is called eschatology, meaning the discussion of 'the last things'. The insight of God's steadfast faithfulness is its foundation for eschatological hope, the basis for theology's affirmation that the universe is not a chaos but a cosmos, not simply a world making sense now but always. Of course, Christians additionally believe that this hope has been manifested and guaranteed by the seminal event of the resurrection of Jesus Christ from the dead. In Christian thinking, what is exceptional about the resurrection is that it occurred within history as the sign of the destiny that awaits all people beyond history. Paul wrote 'for as all die in Adam, so will all be made alive in Christ' (1 Corinthians 15:22).

These beliefs express the essential theological basis for eschatological hope, but nevertheless more discussion seems necessary. It is all very well to speak of a destiny beyond death, but can we begin to make sense of how such a claim could even be a coherent possibility? Reflection soon shows that what would have to be involved in the transition from life in this world to life in a world to come would have to be a subtle mixture of continuity and discontinuity. On the one hand, it really must be Abraham, Isaac and Jacob themselves who live again in the kingdom of God and not just new characters given the old names for old time's sake. There must be sufficient continuity to make this true. Yet, on the other hand, there would not be much point in making the patriarchs live again if they were simply going to die again. There must be sufficient discontinuity to ensure that the new life beyond death is not subject to the mortality that dominates this world. As far as one can, we must explore how we might understand the way in which these dual criteria could be satisfied. We shall look first at the criterion of continuity.

In much Christian thinking over the centuries, the carrier of continuity between this world and the next has been thought of as the human soul, frequently conceived in a platonic fashion as the detachable spiritual part of a person, released at death from its entrapment in the fleshly body. I have already given reasons for rejecting such a dualist view of human nature, arguing instead in favour of seeing human beings as psychosomatic unities. Yet if that is the case, have we not then lost the possibility of the soul being the carrier of continuity? I do not think so, but certainly the soul will have to be reconceptualised. Whatever it might be, presumably it is 'the real me', the carrier of the essence of my individual personhood. It is almost as difficult to think about the nature of this 'real me' within this life as it might be to think of it beyond death. What makes a bald and elderly academic the same person as the schoolboy with a shock of black hair in the photograph of many years ago? It is tempting to think that it is material continuity, but this is an illusion. The atoms that make up our bodies are changing all the time, through wear and tear, eating and drinking. I am atomically distinct from that schoolboy. What links us together is not matter itself but the continuously developing, almost infinitely complex, 'information-bearing pattern' carried at any one time by the matter that then makes up my body. This 'pattern' must be rich enough to incorporate my memories as they accumulate and my character as it forms, together with all else that constitutes me as a person. Clearly there is contact here with the motivated speculations tentatively explored in chapter 3, concerning how some immensely generalised concept of 'information' might offer a glimmer of understanding human nature in terms of a philosophy of dual-aspect monism.

I believe that it is this information-bearing pattern that is the human soul. The idea is, in many ways, an antique one making a fresh appearance in modern dress. Thomas Aquinas, following Aristotle, thought of the soul as being the 'form' (animating principle) of the body and there is clearly some kinship between the idea of form and the idea of information-bearing pattern. Of course, there are differences also. The modern concept has a more dynamic character, corresponding to the growth of personality. It is also more relational, for the pattern that is me is not simply contained within my skin but it must include also the network of personal relationships that do so much to form and sustain me as a person. All these aspects of the soul are beyond present ability to articulate with any adequacy, but they do not seem incoherent ways in which to shape our thinking. We are playing with the toys of thought, but it is play with a serious intention.

The pattern that is me will decompose at my death with the decay of my body, but it does not seem incoherent to believe that God will not allow that pattern to be lost but will preserve it in the divine memory. That of itself would not amount to a life beyond death, for it is surely intrinsic to true humanity to be embodied and this status would need to be restored. We are not apprentice angels. The actual realisation of human life beyond death will therefore require some form of reimbodiment of the preserved pattern of the soul in some new environment of God's choosing. Hence the true fulfilment of Christian eschatological hope is not spiritual survival but final resurrection.

Talk of a new environment raises the issue of discontinuity. The 'matter' of the world to come will have to be different from the matter of this world. It seems a coherent belief that God can bring this about by endowing 'matter' with such strong self-organising principles that it will not be subject to the thermodynamic drift to disorder which is the source of mortality in the present age. Yet considering that possibility immediately raises the question of why, if the new creation is to be free from the death and decay, God did not bring such a world into being from the start? Why bother with the old creation if the new creation is going to be so much better? I believe that the answer lies in the recognition that creation is intrinsically a two-step process. First must come the old creation, existing at some distance from the veiled presence of its Creator so that creatures have the freedom to be themselves and to make themselves, without being overwhelmed by the naked presence of infinite Reality. That world is an evolving world in which the death of one generation is the necessary cost of the new life of the next. Yet, the eventual divine purpose is to draw all creatures into a more intimate and freely embraced relationship with their Creator. The realisation of that purpose will be the coming to be of the new creation in which the divine presence is no longer veiled from view but progressively revealed. This present world contains sacraments, covenanted occasions in which the veil is somewhat thinned. The new creation will be wholly sacramental, totally suffused with the divine presence and so released from bondage to mortality by the presence of the divine eternal life.

The new creation is not to be thought of as a second creation *ex nihilo*, starting from scratch as it were, but it arises *ex vetere*, as the redeemed transformation of the old creation, which is to be valued precisely because of its essential role in the Creator's purposes. God has an appropriate destiny for all creatures and not simply for humanity alone (cf. Colossians 1:20 where Christ is said to reconcile all *things* to God). The destinies of human beings and the destiny of the universe itself form together a single destiny in the faithful purposes of the Creator. For the Christian, the new creation has, in fact, already begun to come into being with the seminal event of Christ's resurrection. The tomb was empty because the matter of his dead corpse had been transformed into the 'matter' of his risen and glorified body. The old and new creations now exist 'alongside' each other in different dimensions (an idea perhaps easier for mathematicians than for most to visualise) and the resurrection appearances could be understood as temporary intersections of these two worlds.

Of course, these are deeply mysterious as well as deeply hopeful matters. Already in 1 Corinthians 15 we find Paul wrestling with issues of eschatological continuity and discontinuity when he says that 'flesh and blood cannot inherit the kingdom of God, nor does the perishable inherit the imperishable . . . we will all be changed' (vv. 50 and 51). Much must remain unclear and unknown to us in this life, but I believe that these tentative explorations are sufficient to show that the eschatological hope of a destiny beyond death is not an incoherent expectation when held in the light of belief in the everlasting faithfulness of the Creator. Several things, however, still remain to be said.

I have said that it is intrinsic to human beings to be embodied, and it also seems clear to me that we are intrinsically temporal beings. This means that there will be 'time' in the world to come as well as 'matter'. Of course the 'time' of that world will be distinct from the time of this world and not merely the latter's prolongation. According to this view, human destiny is not to share in divine eternity through a timeless moment of illumination but to participate in an unending process of divine self-revelation. Finite beings cannot take in the Infinite at a glance. The God who is patient and subtle and content to act in this world through unfolding continuous creation will surely act with similar loving and persistent patience in the life of the world to come. This salvific process will include, as part of its divinely willed fulfilment, purgation and judgement, both hopeful words as they indicate a necessary cleansing engagement with the spiritual realities of individual lives. Judgement involves seeing ourselves as we really are. There will still be dross needing to be purged from us and we shall still have sins not yet fully repented. This cleansing process will be followed by the unending exploration of the inexhaustible riches of the divine nature as it is progressively and clearly unveiled, which means that the everlasting life of the world to come will be far from boring. I do not think that it will only be those who have made a definite commitment to God in this life who will participate in salvation. The decisions and actions we take in this life are certainly very important and it is spiritually damaging wilfully to turn from God, but the divine love and mercy are not on limited offer for this life only. There will surely be further opportunities to turn to God in the clearer light of the world to come. But will there be those who, nevertheless, will reject that love and mercy for ever? We do not know, but if there will be, they will be in hell, not because they have been cast there by an angry God who has lost patience with them but (and this is the tragedy of hell) because they have deliberately chosen to be there. Hell is not a place of unending torture, painted red, but of unending boredom, painted grey, because the divine life that is life indeed has been excluded from it by its inhabitants.

Eschatological hope makes an important contribution to theodicy. It does not do so through a simplistic kind of 'pie in the sky when you die' approach, crudely arguing that the joys of heaven will be enough recompense for any degree of earthly suffering. To say that would be a crass response to a deeply perplexing mystery, but that mystery would be deeper still if there were no prospect of life and fulfilment beyond this present world. The ultimate issue is whether the world makes sense, not simply now but always. Weinberg's answer, given from his naturalistic perspective, is No. Only in the light of seeing the world as a two-step creation, destined to final fulfilment, can one confidently answer Yes.

There has been much speculation in the discussion of this section. In many ways, the best answer to eschatological questioning is to say, 'Wait and see'. The discussion has rested on two solid theological foundations, however, the everlasting faithfulness of God and the resurrection of Jesus Christ.

Scientists opposed to religion often assert that religious faith is a matter of believing without evidence, or even believing against contrary evidence. I have already argued that this is a gravely mistaken view and I have asserted that the issue of attaining truth through well-motivated belief is as central to theology as it is to science, even if the kinds of motivations involved are necessarily different in these two truth-seeking enquiries. Embrace of theism does not demand an act of intellectual suicide in accepting irrational beliefs solely on the basis of unquestioning submission to some allegedly infallible and unchallengeable authority.

The possibility of rational and reliable theological knowledge arises from the conviction that God has revealed the divine nature in a number of ways. The earlier section on natural theology already considered one of these. It claimed that the rationally beautiful order of the universe disclosed to scientific enquiry is best understood as reflecting the Mind of its Creator, and that the finely tuned physical fabric of the universe, necessary to enable the emergence of beings able to pursue that enquiry, is best understood as a sign of God's creative Purpose enabling the fertility of cosmic history. It also briefly suggested that the existence of values, both moral and aesthetic can be understood in terms of there being a divine Ground of value. These insights correspond to what theology terms 'general revelation', hints of a divine presence that arise from the consideration and evaluation of general experience. We also noted that, on its own, this approach can lead only to a generic kind of theism and it leaves unaddressed many sig-

23. RR 5; FSU 3; ST 2; ES.

nificant questions, such as whether the Creator enters into any particular relationship with particular creatures. Resources for addressing such questions will have to come from specific acts of divine disclosure, that is to say, from what theology terms 'special revelation'.

All faith traditions look back to certain particular events and certain particular people, believed to have been specially significant loci of the disclosure of sacred Reality and constituting the foundational source of that tradition. For Christianity, special revelation is focused on the history of Israel and the life, death and resurrection of Jesus Christ. The chapter that follows will pay attention to some of these particular motivations for Christian belief. This section simply seeks to set the scene for that more detailed assessment. The first issue to address is raised by the precise specificity of these claims of special revelatory acts. Why should God have chosen the particular people of ancient Israel as the locus of divine self-revelation? Why should a particular first-century wandering teacher and healer be considered to be the definitive expression in human terms of the divine salvific purposes? Questions of this kind point to what is called 'the scandal of particularity'. I believe that their answer lies in the recognition that in the realm of the profoundly personal there is an inescapable dimension of the uniqueness of every authentic experience, quite different from the repeatable character of impersonal experimentation from which science draws its insights. If Albert Einstein had not discovered general relativity, no doubt it would eventually have come to light through the labours of others, but if J. S. Bach had not composed the Mass in B Minor, that great work of art would have been lost to us for ever. In the section on miracle, I said that God does particular things in particular circumstances and that is why personal language is the least misleading way for us to talk about the divine nature. The scandal of particularity does not mean that God has capriciously chosen special favourites to form a secret inner circle of illuminati (in the way the Gnostics believed), but simply that the good news of divine grace, which is God's gift to all people, has to be communicated initially though the insights and experiences of particular people.

The record of these special revelatory events is what theology calls scripture. For the Christian, the Bible is not to be thought of as a divinely dictated textbook with all the ready-made answers laid out in it as a set of infallible propositions that have to be accepted without question. Rather, it is a kind of spiritual laboratory notebook, containing the humanly written accounts of uniquely significant events of divine self-disclosure. These disclosures themselves are an unfolding tale of an increasing recognition of the nature of God and an increasing understanding of the divine will. The Hebrew Bible, which Christians call the Old Testament, was compiled over a period of almost a thousand years. In its oldest strata we find accounts of war and genocide that greatly trouble us as readers today. They are there because, while the early Hebrews grasped that there is only one God who is to be worshipped and obeyed, they had not yet come to realise that these exclusive claims did not imply a religious duty simply to destroy the worshippers of other gods. The later writings of the prophets attained a wider and truer vision, which saw that God has a concern and a purpose for all the peoples of the Earth. Jesus of Nazareth commanded his followers to love their enemies. Any genuine engagement with Christian revelation will have to acknowledge its developmental character, not altogether unlike science's developing grasp of the nature of the physical world. This unfolding character of understanding has continued beyond New Testament times, expressed in doctrinal developments that Christians believe have been guided by the Holy Spirit, as the implications of the foundational witness of scripture have been further explored and understood.

In reading scripture, respect must be paid to the genre of what has been written. A simple flat-footed literalism would be an abuse of scripture, disrespectful of its character. It is a very bad error to mistake poetry for prose. ('My love is like a red, red rose' does not mean that the poet's girlfriend has green leaves and prickles!) Much of special revelation is contained in accounts of historical events and persons, but some of it is conveyed through symbolic stories. This is the real meaning of the word myth, which does not mean an incredible fairy story but a way of expressing truth too deep to be expressed in any other form than story. A good example of this genre is the story of the Fall in Genesis 3. Read literally as an account of a single disastrous ancestral event that brought death into the world, it is clearly incompatible with what we now know about the evolution of hominids. Read symbolically, it conveys profound insight. I understand it in the following way,[24] relating it to the emergence of self-consciousness, our human power to think beyond the present and project our minds far into the past and far into the future. I do not think that even our closest animal cousins possess this ability, but they live in what one may call the near present. The chimpanzee can figure out that throwing up the stick may dislodge the banana, but he does not sit brooding about the fact he will die in some

24. RR 8.

years' time. Human beings do. I think that at the same time that this self-consciousness dawned there was also a dawning consciousness of the presence of God. Our ancestors turned away from the pole of God into the pole of the self, becoming, as Luther said, curved in upon themselves. This was the process of the Fall, and we are all the heirs of it. It has had two consequences. One is the presence of a moral slantedness in human affairs, which often turns a country's liberator into its next tyrant and is the source of meanness, cruelty and dishonesty in everyday life. Human beings are not wholly autonomous beings, for whom fulfilment comes from 'doing it my way', but we are intrinsically heteronomous, needing to acknowledge and receive the grace of our Creator if we are to live truly fulfilling lives. The refusal to acknowledge this creaturely dependence is what the Christian tradition calls sin. The temptation that the serpent whispers in Eve's ear to persuade her to eat the forbidden fruit is that by so doing she will become 'like God', that is, she can become her own god. The other consequence of the Fall was not biological death, which had been in creation for millions of years before human beings, but what one may call mortality, human experience of sadness at the transience of human life. Our ancestors' self-consciousness enabled them to anticipate their deaths, but their turning away from the God who is the only ground for belief in a destiny beyond death deprived them of that hope. This interpretation of Genesis 3 illustrates how one can bring new understanding to the interpretation of scripture through allowing it to interact with other sources of truth. Scripture does not bear a single meaning but, as with all great classic literature, there is a rich depth of meaning awaiting further exploration by truth-seeking readers, through their making use of the insights available to them in their own generation.

One further question might be asked about the Fall. How can we be sure that the life of the world to come will not be threatened by the possibility of a second such disaster? I think that the answer must lie in the belief that, in the brighter light of the new creation, the divine love and power will be so clearly manifested that all who enter into that world will freely and wholeheartedly embrace the divine will, without reserve.

Earlier I spoke of the way in which the discourse of theology has to range over the centuries. As part of this truth-seeking process, the great insights of special revelation given in the past have to find a resonance in the contemporary experiences of individual believers. An important part of the motivation of a Christian believer will lie in the way in which he or she encounters the reality of Christ today, in reflection on scripture, in worship and in obedient service.

CHAPTER FIVE

Motivated Christian Belief

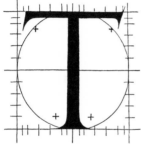HIS short chapter does not claim to set out adequately a wide-ranging and detailed defence of Christian belief. That would require, at the very least, a book for its development.[1] Instead, its intention is simply to give a sketch illustrating some of the lines of argument that need to be pursued in mounting such a defence. Its purpose is to indicate how Christian belief can be seen to find its basis in experience carefully evaluated, and so to rebut allegations that faith is the product of rash and ungrounded speculative thinking or based merely on fideistic assertion. To this end, the discussion will concentrate on two core beliefs that are counterintuitive to secular expectation but central to Christianity, and it will outline motivations for accepting them. The first such belief is in the special significance of Jesus of Nazareth, ex-

1. See SCB for a much more extensive defence of Christian belief.

pressed in the mysterious and exciting claim that deity as well as humanity was truly present in him. If this is true, it follows that Jesus gives us the most direct access possible to understanding the divine nature and purpose, for he is, so to speak, God made known in human terms. An important part of assessing this extraordinary claim will be a careful consideration of the belief that, on the third day after his crucifixion, he was raised from the dead to a life of unending glory. The second core belief is the Trinitarian nature of the Godhead, the fundamental Christian understanding that the unity of the one true God is constituted by the exchange of love between three divine Persons.

The sections addressing these beliefs seek to outline the kinds of motivating evidence that Christians believe support them. The character of the arguments given is exemplary, rather than exhaustive. The aim is not to give a complete treatment, but to say enough to indicate that there are rational motivations for these beliefs to which a bottom-up thinker, open to possibility of the unexpected, should be prepared to give serious consideration. Neither the Christian believer nor the unbeliever has access to indubitable proofs of the kind that it would be wholly irrational to dismiss, but there is the possibility for serious discussion that can be pursued with intellectual integrity and truthful intent.

CHRISTOLOGY[2]

Theological discussion of the true significance of Jesus of Nazareth is called Christology. Assessment of the claim for his unique status must begin by considering what we know

2. SP 8; SCB 5-7; S as T 6; GH 6; ER 4; TCS 6.

about Jesus' life in first-century Palestine. The four Gospels are the principal sources available to us for this purpose— representing, in fact, a larger body of material than is available in relation to many other famous figures in the ancient world. Much argument has been made about what degree of historical reliability can be attributed to the Gospels. Rather than attempting here a detailed account of this involved and controversial discussion, let me say that I believe that a candid reading of the gospels will lead to the recognition that the inspiration behind them derives from the words and deeds of a historical figure with a highly original and challenging personality, and that they do not have the appearance of being a bunch of made-up tales about a shadowy and largely unknown figure, concocted by a variety of different early Christian communities. Serious scholarly study supports this conclusion.[3] I think that one indication of the truthful intent of the gospel writers can be seen in the way in which they record difficult things that Jesus said—for example, 'let the dead bury their own dead' (Matthew 8:22), a truly shocking statement to make in the ancient world where the duty of proper burial was a sacred filial obligation.

But neither are the Gospels scientific history in the modern manner, with its concern with scrupulous attention to minute detail. In the ancient world, those writing about persons of significance were concerned to convey the essential character of their lives, without striving for absolute consistency in minor matters. It is instructive to compare the three accounts given in Acts (chapters 9; 22; and 26) by the same

3. See, for example, R. Bauckham, *Jesus and the Eyewitnesses*, Eerdmans, 2006.

author, which all tell the same essential story of how Paul's life was transformed on the road to Damascus by an encounter with the risen Christ, but which differ in minor details, such as how many saw the heavenly light and who fell to the ground. These minor discrepancies clearly did not trouble Luke.

From the outset, it is necessary to distinguish between Matthew, Mark and Luke (the Synoptic Gospels, so called because they share a large degree of common perspective), and John. The Jesus we encounter in the Synoptics is recognisably rooted in first-century Palestine, giving much of his teaching in striking but homely parables that are set in everyday life, and focussing expectation on the coming of the Kingdom of God. The perspective of John's Gospel seems to be different in character, as much eternal as temporal. Jesus speaks in timeless tones, often focusing the discourse on himself. He asserts an astonishing personal status (for example, 'I am the way, the truth and the life'; 14:6) and claims a uniquely intimate relationship with God his heavenly Father ('Whoever has seen me has seen the Father'; 14:9). It seems to me to be likely that, quite in accordance with acceptable historiographical convention in the ancient world, much of this kind of discourse arose from later long reflection by the writer on the deep significance of Jesus, rather than being a report of words literally spoken by him in his lifetime. Nevertheless, powerful personal claims are not wholly absent from the Synoptics either. 'No one knows the Father except the Son and anyone to whom the Son chooses to reveal him. Come to me, all you that are weary and are carrying heavy burdens, and I will give you rest' (Matthew 11:27-28). When Jesus speaks, what he says is not presented simply as a message conveyed from God (as in the case of the Hebrew prophets) but as strong assertions made apparently on his own authority. For example, 'It was said to those of ancient times [i.e. to Moses on the holy mountain] . . . but I say to you' (six times in the Sermon on the Mount; Matthew 5). People were scandalised by his claiming to exercise the divine prerogative to forgive sins (for example, Luke 5:21) All four Gospels also report many powerful acts of healing performed by Jesus. These accounts are integral to the text and they cannot be excised without doing violence to it. For example, it is clear that there was a bitter controversy with the synagogue authorities about the appropriateness of healing on the Sabbath, which could not have arisen unless there had been such healings.

Of course, Jesus did not go around Palestine saying 'I am God'. No sane human being could do that, and it is an important part of Christian belief about Jesus that he was indeed fully human, even if there was more to say about him than simply that. There is something uniquely challenging about Jesus of Nazareth that resists assimilating him simply to conventional Jewish categories for unusual people, such as teacher, healer or prophet. Exploring what that extra might be leads to the need to consider what truth might lie in the claims made by his earliest followers that he was raised from the dead, three days after his execution. The resurrection poses the central question which must be answered if one is to make a motivated judgement on the status of Jesus and the truth of the Incarnation.

However powerful and successful Jesus' public life might have seemed to be, in the end it appeared to have ended in final failure. He is arrested and his closest followers desert him and run away. He suffers the painful and shameful fate of execution by crucifixion, something that a pious first-century

Jew would have seen as a sure sign of God's rejection, since Deuteronomy (21:23) says 'anyone hung on a tree is under God's curse'. Matthew (27:46) and Mark (15:34) are frank enough to report that as Jesus hung dying on the cross he cried out, 'My God, my God, why have you forsaken me?' I believe that if the story of Jesus ended at Golgotha we would never have heard of him. He would have disappeared from historical memory as yet another first-century messianic pretender whose claims were shown in the end to be delusory. Yet we have all heard of Jesus. *Something* must have happened that continued his story.

All the New Testament writers are clearly convinced that this 'something' was the resurrection. The New Testament does not make sense unless one recognises that this conviction underlies it. It is important to understand the magnitude of the claim being made. Resurrection is quite different from resuscitation. There are stories in the New Testament, and elsewhere, of persons, such as Lazarus, being restored to life after apparent death, but they receive only a temporary reprieve from mortality. They will certainly die again in due course and then they will remain dead. The claim being made about Jesus is that his Resurrection is an everlasting victory over mortality. He was raised to a life of unending glory.

The New Testament presents two principal lines of evidence in support of this astonishing assertion, as counterintuitive in the first century as it may seem to us in the twenty-first century. (There were Jews in the first century who expected a general resurrection at the end of history, but no one expected an individual to be resurrected within history.) One line of evidence is the sequence of appearance stories recounting how the risen Christ met with his disciples. The earliest account we have is given by Paul in his first letter to the Corinthians (15:3–8). He was writing about the year 55 (the crucifixion was either 30 or 33) and he simply gives a list of such appearances, ending with his own dramatic encounter on the road to Damascus. In giving this list, Paul says that he is passing on what he 'in turn had received'. The natural interpretation of this phrase is that it implies that he was recalling teaching which he had been given following his conversion. This would take the testimony back to within a very few years of the events themselves. The antiquity of the testimony receives some confirmation from features of the text, such as Peter being referred to by his Aramaic name, Cephas. Paul is clearly appealing to what he regards as evidence—many of the witnesses are said still to have been alive at the time of his writing—but his concise listing of names alone does not offer any description of what these experiences were actually like. For that we have to turn to the Gospels. At first sight it might seem that we are faced with a bewildering confusion, consisting of a variety of different stories, some set in Jerusalem and some in Galilee. Could this variety not simply reflect the fact that we are being presented with a bunch of made-up tales, originating in the pious imaginations of a number of different communities? I do not think so, for there is a recurrent theme, hardly likely to have arisen with such consistency from a gaggle of independent sources, namely that it was initially difficult to recognise the risen Christ. For example, Mary Magdalene took him to be the gardener (John 20:15), the couple on the road to Emmaus only recognised at the end of the journey who their companion had been (Luke 24:31); Matthew even tells us that on a Galilean hillside 'some doubted' who it was (28:17). It seems to me that this unexpected feature is more likely to be a historical reminiscence of the character of actual encounters, rather than a fortuitous coincidence in a set of independent confabulations.

The second line of evidence is the story of the empty tomb, testified to in all four Gospels, with only minor variations of detail between the accounts. It is certainly true that Paul does not mention the empty tomb, but the occasional character of his letters means that they should not be treated as if they were systematic attempts to list all the arguments in favour of his Christian convictions. What is certain is that Paul believed that Jesus was alive, something that a first-century Jew, taking a psychosomatic view of the nature of human personality, would have been very unlikely indeed to have believed if he also thought that there was a corpse lying mouldering in a tomb. But was there in fact an identifiable tomb at all? It was certainly normal Roman custom to inter the bodies of executed malefactors in an anonymous common grave, but there is also archaeological evidence which shows that this was not invariably the case. The story of the tomb gains credence from the fact that it is Joseph of Arimathea who is said to have provided it. Joseph is an otherwise unknown character of no manifest importance in the early Christian community and the most likely reason that he is assigned this honourable role is that he actually performed it. But the most persuasive argument in favour of the authenticity of the empty tomb story is that it is women who make the discovery. In the ancient world, women were not considered to be capable of being reliable witnesses in a court of law and anyone making up a tale would surely have assigned the central role to men. It is also significant that in the arguments that soon arose between the Jewish and early Christian communities, it is accepted by both sides that there was an empty tomb. They only differ in their explanations of this fact: deceit by disciples who stole the body or resurrection? We can trace this conflict back into the first century (see Matthew 28:11–15). I find it impossible to believe that those who were later martyred for their Christian beliefs went to their deaths knowing that they had been perpetrators of a hoax.

Much more could be said about the detail of these lines of evidence, but enough has been presented to indicate that belief in the resurrection as an actual event, taking place in history though transcending normal historical expectation, is not without serious motivation to which a truth-seeking bottom-up thinker should be prepared to give careful consideration. However, how the evidence is weighed will depend upon the wider context within which it is evaluated. Those with an unrevisable prior commitment to the conviction that what usually happens is what always happens will be driven to dismiss all the evidence as legendary. Yet no one prepared to be open to the possible unexpected strangeness of reality need be forced to that conclusion. In discussing miracle in chapter 4, I emphasised that it is entirely rational from a theistic point of view to believe that in unprecedented circumstances God may do unprecedented things.

At this point in the discussion we encounter an inescapable circularity, similar to the hermeneutic circularity connecting theory and experiment that we identified in science. If Jesus was simply another prophetic figure, or even no more than a messianic pretender, then no doubt after his execution he stayed dead. Yet if Jesus was more than that, the Son of God in some unique sense, then it is a coherent possibility that he was raised from the dead as a sign of that unique status.

Conversely, if Jesus was resurrected, that surely indicates that there was something uniquely significant about him. Moreover, for Christians part of the uniqueness of the Resurrection lies in the fact that it occurred within history as the anticipation of a destiny that God purposes for all other human beings beyond history (1 Corinthians 15:22: 'For as all die in Adam, so all will be made alive in Christ'). From this point of view what is radically exceptional about the resurrection is its timing. The existence of circularity in the argument about Jesus' status and his resurrection means that one cannot claim for it absolute rational certainty, but at the same time, as in science, the presence of circularity does not at all imply that the belief is unreasonable and unmotivated. Christians feel reinforced in their conviction by the fact that the worshipping experience of the Church down the centuries, particularly in the celebration of the Eucharist, has always been of a kind that leads it to speak of Christ as its living Lord in the present, and not just as an impressive founder figure in the past.

Belief in the truth of the resurrection is one reason why the New Testament writers, when they seek to express their experience of Christ, are driven time and again to use about him divine-sounding language, despite their being monotheistic Jews with certain knowledge that he was a man recently alive in Palestine. The earliest Christian confession seems to have been 'Jesus is Lord', quoted by Paul (Romans 10:9; 1 Corinthians 12:3), who calls Jesus 'Lord' more than two hundred times in his letters. This confession of the Lordship of Christ clearly predates the Greek-speaking Pauline Church, as one can see from the fact that 1 Corinthians (16:22) quotes the phrase *maran atha*, O Lord come, in the Aramaic language used by the very earliest Christian communities. While the

Greek *kyrios* could amount to no more than a common courtesy in ordinary speech (rather like the English usage of 'sir'), readers encountering the word in the context of Paul's theological writing would irresistibly be reminded that 'Lord' was the word that pious Jews used in place of the unutterable divine name. This use of divine language about Jesus was motivated not only by belief in the resurrection but also by the experience of transforming new life that these first Christians believed they had received from the risen Christ.

In the New Testament, the issue of how the Lordship of Christ relates to the fundamental Jewish insight of the Lordship of the one true God is left unresolved. For the writers the facts of experience were everything and theory-making was postponed. In scientific terms the situation corresponded to what the physicists call 'phenomenology', the record of significant events that are still awaiting incorporation into an overall theoretical understanding. Later generations of Christians could not let the matter rest there. Eventually this led to the theological theory-making expressed in the orthodox doctrine that in Christ there are two natures, the human and the divine.[4] If Jesus was not truly human he would have no significant relationship to us, but if he was simply human he could be no more than an inspiring exemplary figure. What human beings need is not just an example, but the power to follow that example. For Jesus to be the source of the divine grace for others the life of God had truly to be present in him.

The doctrine of the Incarnation offers theology a profound insight as it wrestles with the problem of evil and suf-

4. See, for example, G. O'Collins, *Christology*, Oxford University Press, 1995.

fering. The Christian God does not just look down with compassion on the travail of creation, viewed from the invulnerability of heaven, but in the cross of Christ we see that God is truly 'a fellow-sufferer who understands' (in Whitehead's phrase), knowing that suffering from the inside, so to speak. This moving insight seems to me to meet the problem of suffering at the deepest possible level of divine response and theological insight.

Much more could be said, and has been said, but even this short sketch may serve to indicate something of how the counterintuitive belief in the Incarnation was motivated by evidence and experience and not by unbridled metaphysical speculation. The idea of a divine sharing in human life and triumph over death is certainly a powerful myth. However, as a Christian I believe that it is more than that, for it is an *enacted* myth.

TRINITY[5]

The first Christians were Jews who knew Israel's Lord, the one true God and the Creator of the world. Nevertheless, as we have seen, they also believed that they had encountered deity in the risen Christ and they confessed Jesus as Lord also. They were also aware of a divine power at work within them, which they called the Holy Spirit, or the Spirit of God or the Spirit of Christ (all three terms are used in the same verse in Romans 8:9). Succeeding generations of Christians have continued to share in this triune character of encounter with the one true God, met with as Father (God above: the almighty Creator and source of being), Son (God alongside: deity

5. ST 3–4; ER 5.

made known in human terms) and Holy Spirit (God within: at work in the human heart, transforming and empowering), without denying their fundamental conviction that there is a single divine Will and Purpose at work in creation. This complex experience is the motivating evidence that eventually led the Church to the formulation of the doctrine of the Trinity. It arose from the necessity to hold fast both to a triune encounter with divinity and to the conviction of the essential unity of the Godhead. The doctrine seeks to do so through understanding that the divine unity is constituted by the eternal exchange of love between three divine Persons.

Once again we see that a counterintuitive Christian doctrine has not arisen from ungrounded metaphysical speculation but from the 'phenomenology' of scriptural witness and worshipping experience. There is an irreducible mystery in infinite deity that means that theological theory-making by finite human beings will never attain perfection but must be content with reaching a modest degree of understanding. In its Trinitarian exploration, the Church was able to identify two extreme positions which it considered would not do justice to its experience and which therefore constituted boundaries within which thought about the triune God had to be contained. One of these extremes was modalism, the idea that Father, Son and Spirit are simply labels for three different aspects of an encounter with what is really an undifferentiated divine unity. Modalism was judged inadequate because it failed to do justice to the distinctive character of the manifestations of the divine Persons. A key counterexample, often appealed to by the Church Fathers, was the baptism of Christ, in which the heavenly voice of the Father proclaimed the beloved character of the Son, on whom the Spirit descended in the form of a dove. The distinctiveness of the roles displayed in this symbolically highly significant event spoke against a modalist assimilation of the Persons. At the same time, a kind of tritheism, in which Father, Son and Spirit are conceived as being three independent members of a mini-pantheon, failed to do justice to the unity of divine Will and Purpose. In wrestling with the task of finding an acceptable middle position between these two extremes, the Church was led to the concept of the three divine Persons whose essential unity is constituted by the eternal exchange of love between them, a mutual indwelling that the Fathers called perichoresis. The struggle to articulate this idea led the Church to have recourse to using and developing the language of Greek philosophy about substance and personhood, in a fashion quite different in character to the phenomenological discourse of the New Testament. This is not the place to attempt a summary of the subtleties involved in this difficult task. No doubt there was a speculative character to some of it, of a kind not dissimilar to the way in which scientists sometimes have to speculate about a difficult and unclear problem, such as in the search for a Grand Unified Theory in particle physics. Nevertheless, it was undeniable primary experience that drove the Trinitarian theologians to their enterprise.

A successful scientific theory very often yields a bonus in the form of offering understanding going beyond the phenomena that motivated its original formulation. Trinitarian theology can claim an analogous gain by offering a deeper understanding of the fundamental Christian belief that 'God is love' (1 John 4:16). This love is surely not only manifested externally in the Creator's love for creation, but it must also be manifested internally in the life of the Godhead. An unnuanced monotheism could only express the internal character of divine love in terms of a ceaseless self-regard, in the manner of the God of Aristotle lost in isolated narcissistic self-admiration. This unsatisfactorily static picture of divine love is replaced in Christian thinking by the eternal dynamic interchange of love between the divine Persons, thereby offering a much more profound understanding.

Other Faiths

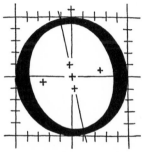NE of the most striking differences between science and religion is that science, once the dust has settled on an issue, succeeds in attaining universal acceptance of its new understanding, while the world of religion is characterised by the continuing existence of very diverse faith communities. Ask suitable persons in London, Delhi or Tokyo what matter is made of, and in all three cities you will receive the same reply, 'quarks, gluons and electrons'. Ask persons in the same three cities what is the nature of ultimate Reality, and you are likely to receive three quite different answers. Does not this fact imply that science gives us real knowledge, but the best that religion can offer amounts to no more than a variety of culturally shaped opinions? I do not think so, but I must acknowledge that one of the most pressing problems facing theology in the

twenty-first century is the challenge presented by the diversity of the world faiths.[1]

Today it is no longer possible to dismiss people of other faiths as strange persons in far-away countries who believe peculiar things. They are our neighbours, living down the street, and we can see the spiritual integrity of their lives. It cannot be the case that we know all the truth and that they, in their ignorance, need to submit in every respect to our superior understanding and have nothing to say to us in return. What is clear is that all the world faith traditions are all testifying to a realm of human experience that can be characterised as encounter with sacred Reality. The problem is that the traditions seem to have such different and incompatible things to report about the nature of this encounter. The disagreements do not relate only to core beliefs, such as the Christian conviction of the unique status of Jesus Christ or the Islamic conviction of the supreme authority of the Qur'an. They also relate to general metaphysical understandings. What is the nature of the human person? The three Abrahamic faiths, Judaism, Christianity and Islam, all say that the individual person is of unique and abiding significance. Hindus believe that the person is recycled through reincarnation, while Buddhists believe that the personal self is ultimately an illusion from which to seek release. These are not three sets of people saying the same thing, expressed in culturally different ways. They are three sets of people in fundamental disagreement, saying three quite different things.

The attempt to wrestle with the problems presented by

1. SCB 10; S as T 5; ER 7.

the diversity of the world faiths will require long and painful dialogue between them. I believe that the complexity of the issues is such that this dialogue is likely to be a task for the third millennium and not just the twenty-first century. I can only offer three simple observations.

The first is that, as the dialogue begins, it should start with questions that are serious but not from the outset disturbingly confrontational in relation to the traditions' core beliefs. If some such caution is not observed, the encounter may rapidly become too threatening and barriers will immediately be set up, with the result that little progress will be made. The important, but not immediately divisive, issue of how the faiths understand their traditional beliefs to relate to the insights of modern science provides an example of a topic that can yield constructive interaction without provoking immediate frustrating defensiveness. In fact, some modest degree of progress has already been made through meetings of this kind, but much more work remains to be done.[2]

The second point is that I do not believe that the solution will lie in the direction of constructing a lowest-common-denominator 'world religion', formed by collecting together what the faiths are found to have in common. There certainly are such commonalities—for example, testimony to the mystical experience of union with the One or the All, and the acknowledgement of the value of compassion—but on their own these amount to something that falls far short of the spiritual depth and vibrancy to be found in each of the individual tra-

2. W. M. Richardson, R. J. Russell, P. Clayton and K. Wegter-McNelly (eds), *Science and the Spiritual Quest: New Essays by Leading Scientists*, Routledge, 2001.

ditions. Once defensive reserve has been broken down, the faiths will have to meet each other in a manner that acknowledges and respects the specificities of insight and experience on which each tradition is based. Mutual incompatibilities will have to be faced and not smoothed over as if they were not that important. I do not think that I would be serving my brothers and sisters in other faiths if I sought to disguise my fundamental conviction of the unique significance of Jesus Christ, even though I know that they do not share that belief with me. It is precisely this need to be true to what we believe that we have been given that will make the interfaith dialogue long and painful.

My final observation is that, from the Christian perspective, I believe that the way to understand the spiritual authenticity present in the other faith communities is in terms of the working of the Holy Spirit. There is a Christian tradition which understands much of the working of the Spirit to be veiled, hidden from explicit view until its final manifestation in the fulfilment of the coming of the new creation.[3] In the Gospel of John (15:13), the Spirit is called 'the Spirit of truth' and is, I believe, at work in all truth-seeking communities. The reticent veiling of the Spirit means that the fruits of the Spirit are not confined to those who know him by name. I offer this thought, not as an imperialistic attempt to appropriate to Christianity the spiritual authenticity of other faith traditions, but simply as a way for the Christian to understand and honour that authenticity.

---

3. V. Lossky, *The Mystical Theology of the Eastern Church*, James Clarke, 1957, ch. 8.

Index

Abraham, 102–4
Active information, 47, 85–86
Acts, Book of, 118–19
Adam, 113–15, 125
Aesthetic values, 77–78
Agency. *See* Free choice
Altruism, 61
Anaxagoras, 66
Anaximander, 66
Angels, 106
Anthropic Principle, 54–56, 74
Anthropology, 87
Antimatter, 5
Apophaticism, 12, 20
Aquinas, Thomas: on God, 64, 88–89, 98; on human free choice, 64; influence of, 11, 14; on soul, 105
Aristotle, 16–17, 28, 105, 130
Astronomy, 27–29, 31–32
Atoms, 2, 35, 49
Augustine, 14, 28

Bach, J. S., 14, 111
Baptism of Jesus Christ, 128–29
Barbour, Ian, xii, 20–25
Barrow, J., 54n11
Bartholomew, D. J., 88n14
Bauckham, R., 118n3
Beethoven, Ludwig van, 14
Belief in God in an Age of Science

(Polkinghorne), 16n8, 20n9, 47n8, 70n1, 84n9, 86n11
Bellarmine, Cardinal, 28
Bénard convection, 45
Beyond Science (Polkinghorne), 2n1
Bible. *See* New Testament; Scripture; *and specific books of the Bible*
Big bang cosmology, 47–49, 51–52
Block universe, 62–64, 98–99
Bohm, David, 37–38, 40
Bohr, Niels, 37–40
Bottom-up thinking, xi, 17–20, 88
Brain, 61–62, 65, 67. *See also* Consciousness
British Association for the Advancement of Science, 30
Brooke, John Hedley, 26
Butterfly effect, 39

Calvin, John, 14
Cancer, 61
Carbon, 54–56, 74, 102
Causality: and chaos theory, 39–42; divine primary causality, 88–89; and Divine Providence, 89–91; and metaphysics, 38, 40–41, 84–85; nature of, 11; and quantum physics, 35–42; and science, 34–42; temporality compared with, 64; top-down causality, 47, 85, 88

CBR (cosmic background radiation), 49

Chance: definition of, 57; evolution and interaction of Necessity and, 57–62, 80–81, 82

Chaos, edge of, 59–60, 61, 80–81, 82

Chaos theory, 39–42

Chemistry, 7

Chinese, 31, 32

Christianity: and Christology, 116–27; clarification of belief and correction of heresy in, 15–16; and diversity of world faiths, 131–34; and Divine Providence, 84–91; and eschatology, 103, 106; and eternal punishment for sins, 16; and evolution, 20–21, 30; and fundamentalists, 20–21, 30; and God of love, 80, 129; and Holy Spirit, 113, 127–29, 134; and Lordship of Christ, 125–26; and miracles, 95–98; and new creation, 106–7, 115, 134; and prayer, 91–94; and resurrection of Christ, 25, 95–96, 103, 107, 109, 120–26; and revelation, 76, 112–13; and Scripture, 112; and sin, 114; and slavery, 16; and Trinity, 15, 77, 99, 117, 127–30; and two natures of Christ, 15, 17, 97, 116–17, 120, 126–27; and world as God's creation, 31–32. *See also* Jesus Christ; Scripture; Theology

Christology, 116–27. *See also* Christianity; Jesus Christ

Clayton, P., 85*n*10, 87*n*12, 133*n*2

Commitment, 9–11, 13, 19, 71

Compassion, 133

Complex systems, 44–47

Computerised models of logical networks, 45–46

Conflict between science and theology, 20–21

Consciousness, 65–68, 113–14

Consistency of God, 96–97

Convergence, 59

Conway Morris, Simon, 58–59

Copenhagen interpretation of quantum theory, 37–39

Copernicus, 27

Corinthians, First Letter to, 103, 107, 122, 125

Correspondence principles, 8

Cosmic background radiation (CBR), 49

Cosmological constant, 54

Cosmology, 3, 47–57, 64–65. *See also* Universe

Councils of the church, 15

Creatio continua (unfolding creative process), 79–80, 99

Creatio ex nihilo (creation out of nothing), 79

Creation, 29–32, 73–86, 88–90, 128, 129. *See also* New creation

Critical realism, 10–11, 15

Crucifixion, 120–21, 122, 127

Dark matter and dark energy, 53–54

Darwin, Charles, 3, 4, 29–30, 79. *See also* Evolution

De la Mettrie, Julien Offray, 32

Death, 57, 102–6, 114, 120–21, 125

Deism, 24–25, 76, 80, 84, 90, 91–92. *See also* God

Deity. *See* God

Demiurge, 31

Determinism, 64, 88

Deuteronomy, Book of, 121

Developmental stance in theology, 25

Dialogue between science and theology, 20, 22, 69–70. *See also* Science; Theology

Dialogue Concerning the Two Chief World Systems (Galileo), 27

Dirac, Paul, 5, 37

Disasters, 82

Diseases, 82

Dissipative systems, 44–46

Divine Providence, 84–91. *See also* God

DNA, 3

Dual-aspect monism, 66, 105

Dualism, 65–66

Earth: age of, 20; and Copernican system, 27, 28; history of life on, 20, 33, 55–56, 59, 78, 81; as life-bearing planet, 55–56

Edge of chaos, 59–60, 61, 80–81, 82

Einstein, Albert, 43–44, 53–54, 111

Electricity, 97–98

Electrons, 2, 5, 6–7, 17, 41

Encountering Scripture (Polkinghorne), 110*n*23

*End of the World and the Ends of God,
The* (Polkinghorne and Welker),
102*n*21
Enlightenment, 10
Eschatology, 102–10
Eternity, 108
Eucharist, 125
Eve, 113–15
Evil, 81–84, 126–27. *See also* Sins
Evolution: and altruism, 61; contextual
factors in, 60–61; and convergence,
59; and *creatio continua* (unfolding
creative process), 79–80; creation,
evil and, 78–83; Darwin's theory of,
3, 4, 29–30, 79; and fundamentalists,
20–21, 30; and genetics, 61–62; and
interaction of Chance and Necessity, 57–62, 80–81, 82; of mathematical abilities, 60–61; Aubrey
Moore on, 29–30, 79–80; and
natural selection, 3, 29, 57–59; and
need for survival, 34; and possibilities at the edge of chaos, 59–60, 61,
80–81; processes of, 57–62; theistic
evolution, 83; and theology, 20–22,
29–30, 78–83; of universe, 58
Evolution, theory of, 21–22
Evolutionary biology, 3, 4–5
Experiments. *See* Science
Exploring Reality (Polkinghorne), 33*n*1,
61*n*14, 62*n*15, 65*n*16, 81*n*5, 98*n*19,
102*n*21, 117*n*2, 127*n*5, 132*n*1

Faith, 13, 90, 110, 116
Faith, Science and Understanding
(Polkinghorne), 62*n*15, 70*n*1, 84*n*9,
98*n*19, 110*n*23
Faithfulness of God, 102–3, 109
Fall, story of, 113–15
Fatherhood of God, 19, 97, 127–29
Feminist theology, 14
Fideism, 15, 18, 89, 116
Field theories, 34–35
Fractals, 42
Free choice, 64, 81–86, 89–91
Free-process defence, 82, 84
Free-will defence, 81–82, 84
Fundamentalists, 20–21, 30

Galapagos Islands, 4
Galaxies. *See* Stars and galaxies

Galileo, 27–29, 31–32
General relativity theory, 5, 7, 8, 44,
48, 50–51, 54
Genesis, Book of, 113–15
Genetic mutations, 58, 59, 60, 61, 81,
82
Genetics, 3, 29, 61–62
Germs, 61
Gifford Lectures, xi
Gleick, J., 39*n*3
Gluons, 2
Gnostics, 112
God: as almighty, 81; anger of, 109;
Aquinas on, 64, 88–89, 98; as Architect of the universe, 24–25; Aristotle on, 130; awe in experience
of, 17; belief in, 11, 24; consistency
of, 96–97; as Cosmic Architect
or Great Mathematician, 76; as
Cosmic Tyrant, 80, 81, 84; as Creator, 29–32, 73–84, 88–90, 96–97,
100, 110, 129; of deism, 24–25, 76,
80, 84, 90, 91–92; dipolar concept of divine eternity/temporality,
99–101; and Divine Providence,
84–91; faithfulness of, 102–3, 109; as
Father, 19, 97, 127–29; and foundational assumption of metaphysics,
23–24; glory of, 100; grace from,
112, 114; as Grand Puppet Master,
80; as ground of value, 77–78, 110;
in Hebrew scriptures, 100–103, 112,
121; hubris of claiming exact knowledge of, 12; humans created in image
of, 31; judgement by, 19, 108–9; and
kenosis, 83, 101; knowledge of, 12, 17,
20; Lordship of, 126, 127; love and
mercy of, 80, 81, 100, 108–9, 129;
Mind of, 76, 110; and natural processes, 70; and natural theology,
70–78, 110–11; obedience to, 17;
omniscience of, 101; perfection of,
99–100; prayer to, 91–94; proofs of
existence of, 18, 71; revelation from,
12, 15, 110–15; and salvation, 108–9;
and scandal of particularity, 111–12;
of theism, 23, 25, 32, 71, 74–78, 92,
99, 110; theology as reflection on
human encounter with, 12, 13, 14,
17; and theology of nature, 76–77;
and time, 98–101; Trinitarian nature

God (continued)
  of, 15, 77, 99, 117, 127–30; unity of,
  15; Will and Purpose of, 76–77,
  128, 129; worship of, 17. *See also*
  Theology
God of Hope and the End of the World,
  The (Polkinghorne), 65*n*16, 102*n*21,
  117*n*2
God of the gaps, 70, 80
Goedel, Kurt, 19, 71
Gospels. *See* New Testament; *and spe-*
  cific Gospels
Gould, Stephen Jay, 58
Grace, 112, 114
Greeks, 31, 129. *See also* Aristotle
Greene, B., 51*n*10

Healing, 93, 94, 120
Heaven, 108–9
Hebrew Bible, 90–91, 100–103, 112–14,
  121. *See also* Scripture
Heisenberg, Werner, 17, 35, 37–38, 52
Helium, 49
Hell, 16, 109
Heresy, 16
Hindus, 132
History: and Gospels, 118; of science,
  26–32
Holder, R. D., 74*n*2
Holism, 42–47
Holy Spirit, 113, 127–29, 134
Hope, 103, 106–9, 114
Human brain, 61–62, 65, 67
Human genome, 61–62
Human nature: as autonomous and
  heteronomous, 114; and conscious-
  ness, 65–68, 113–14; creation of
  humans in image of God, 31; duality
  of, 65–66; and free choice, 64, 81–
  86, 89–91; and mortality, 102, 114,
  121; and soul, 65–66, 104–6; world
  faiths on, 132
Huxley, Thomas Henry, 30
Hydrogen, 49, 56
Hypothesis. *See* Science

Ideas, 31
Incarnation, 83, 120, 126–27
Independence of science and theology,
  20, 21–22

Information: active information, 47,
  85–86; and soul, 105
Inquisition, 27
Integration between science and the-
  ology, 20, 22–25, 69–115
Interfaith dialogue, 131–34
Isaac, 102–4
Islam, 132
Israelites, 90–91, 102–3, 111

Jacob, 102–4
Jaki, S. L., 31*n*2
Jeremiah, 101
Jesus Christ: appearance stories after
  resurrection of, 121–23; baptism of,
  128–29; and Christology, 116–27;
  crucifixion of, 120–21, 122, 127; on
  destiny beyond death, 102; God's
  relationship with, 119–20; healing
  by, 93, 94, 120; human and divine
  natures of, 15, 17, 97, 116–17, 120,
  126–27; as incarnation of divine
  Word, 83, 120, 126–27; Lordship of,
  125–26, 127; on love, 112; miracles by,
  93, 94, 95, 120; parables of, 119; resurrection of, 25, 95–96, 103, 107, 109,
  120–26; and Sermon on the Mount,
  120; as Son of God, 124–25, 127–29
Jews. *See* Judaism and Jews
John, First Letter of, 129
John, Gospel of, 95, 119, 122, 134
Joseph of Arimathea, 123
Judaism and Jews, 31, 120–21, 123–27,
  132

Kaufmann, Stuart, 45
Kenosis, 83, 101
Kepler, Johannes, 31
Kin altruism, 61
Kingdom of God, 119
Kings, Second Book of, 100–101
Kingsley, Charles, 30, 79
Knowledge: of God, 12, 17, 20; reli-
  gious knowledge as dangerous, 13;
  unity of, 15, 22. *See also* Truth and
  understanding
Kuhn, Thomas, 7–8

Lacugna, C. M., 77*n*3
Laplace, Pierre-Simon, 36, 64

Laser, 7, 92–93
Lazarus, 121
Liberation theology, 14
Life after death, 102–10
Life, history of, 20, 33, 55–56, 59, 78, 81
Light: laser light, 92–93; wave/particle duality of, 6, 17, 35–38, 67, 90
Lonergan, Bernard, 11
Lordship: of Christ, 125–26, 127; of God, 126, 127
Lorenz, Ed, 39
Lossky, V., 134n3
Love: among divine Persons in Trinity, 129–30; of God, 80, 81, 100, 108–9, 129; Jesus Christ on, 112
Lucas, John, 93–94
Luke, Gospel of, 119, 120, 122
Luther, Martin, 114

Mach's Principle, 65
Man the Machine (de la Mettrie), 32
Mandelbrot set, 72
Mark, Gospel of, 93, 94, 102, 119, 121
Mars, 8
Martyrs, 124
Mary Magdalene, 122
Materialism, 23
Mathematics: abstract thought in, 34; and chaos theory, 40; and commitment, 19, 71; evolution of mathematical abilities, 60–61; and field theory, 35; and Goedel, 19, 71; and physics, 72–73; and Poincaré, 39; precise language of, 12
Matthew, Gospel of, 118, 119, 121, 122, 124
Mendel, Gregor, 29
Metaphysics: and causality, 38, 40–41, 84–85; and chaos theory, 40; foundational assumption in, 23–25; and philosophical theology, 24; and quantum physics, 38, 40; and scientism, 23; scientists' dismissal of, 11, 22–23; theistic metaphysics, 22–25; and time, 63–64
Metaquestions, 70–75
Milky Way, 55–56
Miracles, 95–98, 111–12, 120
Modalism, 128
Models: computerised models of

logical networks, 45–46; definition of, 19; in science, 19; in theology, 19–20
Moore, Aubrey, 29–30, 79–80
Moral values, 77–78
Mortality, 102, 114, 121. *See also* Death
Moses, 120
Moving present, 64–65
Multiverse, 74–75
Murphy, N., 87n12
Music, 4, 14, 111
Mysticism, 133
Myth, 113, 127

Nagel, Thomas, 66
Natural selection, 3, 29, 57–59
Natural theology, 15, 70–78, 110–11
Nebuchadnezzar, 101
Necessity: definition of, 57–58; evolution and interaction of Chance and, 57–62, 80–81, 82
Neural structure and neural processing, 61–62, 65, 67
Neuroscience, 67. *See also* Brain
New creation, 106–7, 115, 134
New Testament: appearance stories in, of risen Christ, 121–23; on crucifixion, 121; development and correction in understanding of, 15; on eschatology, 102, 103, 107; on healing and miracles by Jesus, 93–95, 120; historical reliability of Gospels, 118; on Holy Spirit, 127, 134; on incarnation of divine Word, 83, 120; on Lazarus, 121; on Lordship of Christ, 125–26; on relationship between God and Jesus Christ, 119–20; on resurrection of Jesus Christ, 121–26; Synoptic Gospels of, 119. *See also* Jesus Christ; Scripture; *and specific books of the New Testament*
Newton, Isaac, 14, 32, 34, 35, 37, 40, 84
Newtonian physics, 7–8, 14, 16–17, 32, 34, 35, 40, 84, 88
Nicene Creed, xi
Nuclear physics, 19
Nucleogenesis, 50

Obedience to God, 17
Objectivity, 10

O'Collins, G., 126*n*4
Ohm's Law, 97–98
Old Testament. *See* Hebrew Bible;
  Scripture
Omniscience of God, 101
On the Origin of Species (Darwin),
  29–30
One World (Polkinghorne), 2*n*1, 33*n*1
Onnes, Heike Kamerlingh, 97–98
Oord, Thomas Jay, xii
Origen, 94
Owen, Sir Richard, 29

Parables, 119
Parmenides, 62
Paul, 103, 119, 122–25
Peacocke, Arthur R., xii, 57*n*12, 79*n*4,
  88*n*13
Pendulum, 52
Perfection of God, 99–100
Perichoresis, 129
Personal Knowledge (Polanyi), 8–10
Petitionary prayer, 91–94
Phenomenology, 126, 128
Philippians, Letter to, 83
Philosophical theology, 24
Physical reductionism, 23
Physics: and atoms, 2; and Bénard con-
  vection, 45; and causality, 35–40;
  and chaos theory, 39–42; as experi-
  mental science, 3; Grand Unified
  Theory in particle physics, 129;
  Heisenberg's uncertainty principle,
  17, 35, 37–38, 52; and mathematics,
  72–73; and moderately complex sys-
  tems, 44–47; Newtonian physics,
  7–8, 14, 16–17, 32, 34, 35, 40, 84,
  88; nuclear physics, 19; and Ohm's
  Law, 97–98; and pendulum, 52; and
  phenomenology, 126; Planck's fundamental constant in, 42; superposition principle of, 6; and second
  law of thermodynamics, 56–57. *See
  also* Quantum theory and quantum
  physics; Relativity
Planck, Max, 42, 50
Poincaré, Henri, 39
Polanyi, Michael, 8–10, 13, 19
Polkinghorne Reader, The (Polkinghorne), xii, xii*n*7
Post-modernism, and science, 10

Prayer, 91–94
Pre-Socratics, 66
Prigogine, I., 45*n*1
Princeton Theological Seminary, 30
Process theology, 100
Proofs of existence of God, 18, 71
Prophets, 101, 112, 119
Ptolemy, 27, 28
Purpose and Will of God, 76–77, 128,
  129

Quantum chaology, 41–42, 87
Quantum entanglement, 42–44
Quantum pendulum, 52
Quantum Physics and Theology (Polkinghorne), 16*n*8
Quantum Theory (Polkinghorne), 6*n*2,
  42*n*5
Quantum theory and quantum
  physics: Bohm on, 37–38, 40; Bohr
  on, 37–40; and causality, 35–42;
  Copenhagen interpretation of,
  37–39; and cosmology, 50–52; and
  Divine Providence, 86, 87–88; and
  evolution, 60; and Heisenberg's
  uncertainty principle, 17, 35, 37–38,
  52; and intelligibility of subatomic
  world, 71–72; logic of, 16–17; and
  metaphysics, 38, 40; and nature of
  science, 5–8; and quantum chaology,
  41–42, 87; and quantum entanglement, 42–44; and quantum pendulum, 52; and quantum vacuum,
  51–54; and superposition principle,
  6, 17, 37; and wave/particle duality
  of light, 6, 17, 35–38, 67, 90
Quantum vacuum, 51–54
Quarks, 2, 11
Quarks, Chaos and Christianity (Polkinghorne), xii, xii*n*4
Qur'an, 132

Radiation, 49
Rational strategy: bottom-up thinking, xi, 17–20, 88; in science, 16–20;
  in theology, 16–20; top-down thinking, 18, 46, 85, 88
Realism: critical realism, 10–11, 15; scientific realism, 5–11
Reason and Reality (Polkinghorne),
  16*n*8, 33*n*1, 110*n*23, 113*n*24

Reciprocal altruism, 61

Reductionism. *See* Physical reductionism

Reincarnation, 132

Relationality, and science, 42–47

Relativity: and block universe, 62–65; general theory of, 5, 7, 8, 44, 48, 50–51, 54, 111; special relativity, 62–65

Religion. *See* Christianity; God; Jesus Christ; Scripture; Theology

Resurrection: general resurrection at end of history, 121; of Jesus Christ, 25, 95–96, 103, 107, 109, 120–26

Revealed theology, 15. *See also* Revelation

Revelation: and Christianity, 76, 112–13; definition of, 12; general revelation, 110; and Holy Spirit, 113; and natural theology, 15, 70–78, 110–11; questions on, 111; and scandal of particularity, 111–12; and Scripture, 15, 110–15; special revelation, 111–15

Revisionary stance in theology, 25

Richardson, W. M., 133*n*2

Roman Catholic Church, 27–29

Romans, Letter to, 125, 127

Russell, R. J., 85*n*10, 87*n*12, 133*n*2

Sacraments, 106–7

Salvation, 108–9

Scandal of particularity, 111–12

Schroedinger equation, 35–36, 38

Science: background problem in experiments in, 9; and belief in unseen realities, 11; bottom-up thinking in, 17–19; and causality, 34–42; and certain proof, 18–19; circularity connecting theory and experiment in, 4–5, 9, 10–11, 124; and commitment, 9–11, 13, 19; conflict between theology and, 20–21; and consciousness, 65–68; correspondence principles among theories in, 8; and cosmology, 47–57, 64–65; and creation of 'maps' of physical world, 7–8; and critical realism, 10–11, 15; dialogue between theology and generally, 20, 22, 69–70; experimental 'fact' and theoretical 'opinion' in, 1–2; experimenter in, 12, 13; experiments and repeatability of, 3–4; as first-order discipline, 24; and Galileo, 27–29, 31–32; history of, 26–32; hypothesis in, 4–5; independence of theology from, 20, 21–22; insights from, 33–68; integration between theology and, 20, 22–25, 69–115; and intelligibility of universe, 10; and intrinsic unpredictabilities in nature, 35; as linearly progressive discipline, 13–14; long-range and long-term fruitfulness of theory in, 4–5, 129; and metaquestions, 70–75; and miracles, 96; mistaken view of truth of theology versus, 1–2; models in, 19; non-realist account of, 28–29; and objectivity, 10; origins of, 30–31; personal judgment in, 9; Polanyi on, 8–10, 13, 19; post-modern account of, as social construction, 10; prediction from theory in, 5; questions addressed by, 21, 24; rational strategy in, 16–20; realist belief in, 5–11; and relationality and holism, 42–47; relationship of theology and, 20–25; scientism versus, 23; self-defining limitation to impersonal experience in, 3–4, 12, 13–14, 17, 23, 70; strangeness encountered by, 5–6, 11, 17, 18; success of, in quest for understanding, 2–5, 8; and time, 62–65; and top-down thinking, 18, 46, 47, 85, 88; truth-seeking in, 2–15, 110, 131; and unity of natural world, 15. *See also* Evolution; Mathematics; Physics; *and specific scientific disciplines and theories*

Science and Christian Belief (Polkinghorne), xi, xi*n*1, 16*n*8, 65*n*16, 79*n*4, 102*n*21, 116*n*1, 117*n*2, 132*n*1

Science and Creation (Polkinghorne), 16*n*8, 33*n*1, 70*n*1, 79*n*4

Science and Providence (Polkinghorne), 81*n*5, 82*n*6, 84*n*9, 91*n*16, 95*n*18, 98*n*19, 102*n*21, 117*n*2

Science and the Trinity (Polkinghorne), 20*n*9, 24*n*11, 33*n*1, 42*n*4, 102*n*21, 110*n*23, 127*n*5

Science and Theology (Polkinghorne), xii, xii*n*5

Scientific realism, 5–11

Scientism, 23

Scientists as Theologians (Polkinghorne), xii, xii*n*6, 13*n*6, 16*n*8, 20*n*9, 79*n*4, 84*n*9, 117*n*2, 132*n*1

Scripture: Augustine on interpretation of, 28; and Christianity, 112; God as author of, 32; Hebrew Bible, 100–103, 112–14, 121; and revelation, 15, 110–15; symbolic reading of, 113–15. *See also* New Testament

Self-consciousness. *See* Consciousness

Sermon on the Mount, 120

Simultaneity of distant events, 63

Sins, 12, 16, 114, 120. *See also* Evil

Slavery, 16, 90–91

Soul, 65–66, 104–6

Southgate, C., 82*n*7

Space program, 8

Stars and galaxies, 50, 53, 55, 56, 58, 65, 78

Steady-state universe, 78

Stengers, I., 45*n*1

Strange attractor, 39

String theory, 51

Suffering, 81–84, 126–27

Sun, 56, 94

Superconductivity, 7, 98

Superposition principle, 6, 17, 37

Supersymmetry, 53, 54

Synoptic Gospels, 119

Systematic theology, 24

Temple, Frederick, 30

Temporality. *See* Time

Theism, 23, 25, 32, 71, 74–78, 92, 99, 110. *See also* God

Theistic evolution, 83

Theistic metaphysics, 22–25

Theodicy, 91, 109

Theology: and acquisition of motivated belief, 15, 19; and apophaticism, 12, 20; and belief in God, 11; bottom-up thinking in, xi, 17–20, 88; clarification of belief and correction of heresy in, 15–16; and commitment, 13; conflict between science and, 20–21; creation, evolution and evil, 78–83; and critical realism, 15; danger of religious knowledge, 13; and deism, 24–25, 76, 80, 84, 90, 91–92; Developmental stance in, 25; dialogue between science and generally, 20, 22, 69–70; diversity of world faiths, 131–34; and Divine Providence, 84–91; and eschatology, 102–10; and evolution, 20–22, 29–30, 79–83; experiential context in, 14–15; and faith, 13, 90, 110, 116; fideism versus, 15, 18, 89, 116; and Galileo, 27–29, 31–32; and human encounter with God, 12, 13, 14, 17; independence of science from, 20, 21–22; integration between science and, 20, 22–25, 69–115; and intelligibility of the world, 11; and metaquestions, 70–75; and miracles, 95–98, 111–12, 120; mistaken view of truth of science versus, 1–2; models in, 19–20; natural theology, 15, 70–78, 110–11; past and present as equally important in, 14; philosophical theology, 24; and prayer, 91–94; process theology, 100; and proofs of existence of God, 18, 71; questions addressed by, 21; rational strategy in, 16–20; relationship of science and, 20–25; and revelation, 12, 15, 110–15; revisionary stance in, 25; and Scripture and revelation, 110–15; as single integrated discipline, 15; symbol as language of, 12; systematic theology, 24; and theism, 23, 25, 32, 71, 74–78, 92, 99, 110; time and God, 98–101; truth-seeking in, 12–16, 110, 115, 131–32. *See also* Christianity; God; Jesus Christ; Scripture

Theology in the Context of Science (Polkinghorne), 15*n*7, 16*n*8, 206*n*9, 62*n*15, 65*n*16, 98*n*19, 102*n*21, 117*n*2

Theology of nature, 76–77

Theory. *See* Science

Theory: circular nature of interaction of experiment and, 4–5, 9, 10–11; long-range and long-term fruitfulness of scientific theories, 4–5; prediction from scientific theories, 5; theological theory-making, 19–20

Thermodynamics, second law of, 56–57

Thomas Aquinas. *See* Aquinas, Thomas

Time: and block universe, 62–65, 98–99; causality compared with temporality, 64; and God, 98–101; and metaphysics, 63–64; and moving present, 64–65; nature of, 11, 62–65; and present moment, 63; and science, 62–65; and simultaneity of distant events, 63; in world to come, 108–9

Tipler, F., 54*n*11

Top-down causality, 47, 85, 88

Top-down thinking, 18, 46, 47, 85, 88

Tracey, T., 87*n*12

Trinity, 15, 77, 99, 117, 127–30. *See also* God; Holy Spirit; Jesus Christ

Tritheism, 129

Truth and understanding: and diversity of world faiths, 131–34; mistaken view of science versus theology, 1–2; and nature of science, 2–15, 21, 110, 131; and nature of theology, 12–16, 21, 110, 115, 131–32; rational strategy for, 16–20; relationship between science and theology, 20–25; and verisimilitude of science, 8. *See also* Knowledge

Tsunamis, 82–83

Uncertainty principle, 17, 35, 37–38, 52

Understanding. *See* Truth and understanding

Unity of knowledge, 15, 22

Universe: age of, 47, 57, 65; and Anthropic Principle, 54–56, 74; and big bang cosmology, 47–49, 51–52; block universe, 62–65, 98–99; and CBR (cosmic background radiation), 49; cooling of, 48–49; cosmic history of, 3, 47–57, 64–65; dark matter and dark energy in, 53–54; end of, 102; evolution of, 58; flatness of, 48; future of, 56–57; inflationary period of, 48; as isotropic, 48; and moving present, 64–65; and multiverse, 74–75; quantum theory and early universe, 50–52; size of, 55–56; stars and galaxies of, 50, 53, 55, 56, 58, 65, 78; steady-state universe, 78; Weinberg on pointlessness of, 102, 109; WIMPS (weakly interactive massive particles) in, 53

Urban VIII, Pope, 27, 28

Vacuum, 51–54

Verisimilitude of science, 8

Warfield, B. B., 30

Weather system, 39

Wegter-McNelly, K., 133*n*2

Weinberg, Steven, 49*n*9, 102, 109

Welker, M., 102*n*21

Whitehead, A. N., 100, 127

Wigner, Eugene, 73

Wilberforce, Samuel, Bishop of Oxford, 30

Will and Purpose of God, 76–77, 128, 129

William of Ockham, 75

WIMPS (weakly interactive massive particles), 53

Work of Love, The (Polkinghorne), 83*n*8

World religions, 131–34

Worship of God, 17